A METHODOLOGY FOR PROCESSING RAW LIDAR DATA TO SUPPORT URBAN FLOOD MODELLING FRAMEWORK

A METHODOLOGY FOR PROCESSING RAW LIDAR DATA TO SUPPORT URBAN FLOOD MODELLING FRAMEWORK

DISSERTATION

Submitted in fulfillment of the requirements of

the Board for Doctorates of Delft University of Technology

and of the Academic Board of the UNESCO-IHE

Institute for Water Education

for the Degree of DOCTOR

to be defended in public on

Wednesday, 21st March 2012, at 12:30 hours

in Delft, the Netherlands

by

Ahmad Fikri Bin ABDULLAH

Master of Science in Hydroinformatics, UNESCO-IHE, The Netherlands

born in Terengganu, Malaysia

This dissertation has been approved by the supervisor:
Em. Prof. dr. R.K. Price

Composition of Doctoral Committee:

Chairman	Rector Magnificus TU Delft
Vice-Chairman	Rector UNESCO-IHE
Em. Prof. dr. R.K. Price,	UNESCO-IHE, supervisor
Prof. dr. ir. F. H. L. R. Clemens,	TU Delft
Prof. dr. ir. L. C. Rietveld,	TU Delft
Prof. dr. A. Abdul Rahman,	University Teknologi Malaysia, Malaysia
Prof. dr. ir. P. O'Kane,	University College Cork, Ireland
Dr. Z. Vojinovic ,	UNESCO-IHE
Prof. dr. ir. J.B. van Lier,	TU Delft, reserve member

CRC Press/Balkema is an imprint of the Taylor & Francis Group, an informa business

Published by:
CRC Press/Balkema
PO Box 447, 2300 AK Leiden, the Netherlands
e-mail: Pub.NL@taylorandfrancis.com
www.crcpress.com - www.taylorandfrancis.co.uk - www.ba.balkema.nl

ISBN 978-0-415-62475-6 (Taylor & Francis Group)

About the Author

Ahmad Fikri bin Abdullah was born in the state of Terengganu, Malaysia. In 1996 he enrolled to the BSc degree course with a full scholarship from the Public Service Department of Malaysia for 4 years in Geoinformatics (GIS) at the Malaysia University of Technology. He was graduated (with distinction) in 2000. Soon after that, he was hired as a GIS Executive at Geomatika Technology Sdn Bhd and after that as a GIS Manager at Guardian Data Sdn Bhd.

In 2002 he was hired as a tutor for the Department of Biological and Agricultural Engineering, Universiti Putra Malaysia. In 2004 he received a full scholarship from the Ministry for Higher Education of Malaysia for pursuing his study to a Master degree. He pursued his study at the UNESCO-IHE Institute for Water Education, Delft, The Netherlands. After 18 months he received MSc. degree in Hydroinformatics. His MSc. thesis was entitled WebGIS Flood Information System.

In July 2006 he was offered a full scholarship UNESCO-IHE under SWITCH project for PhD degree. In 2008 he received a full scholarship from the Ministry for Higher Education of Malaysia for pursuing his PhD. The period of the scholarship was 5 years. His research was devoted for A Methodology for Processing Raw LiDAR Data to Support Urban Flood Modelling Framework which is presented in this thesis.

List of Publications

Abdullah (2007), WebGIS Flood Information System, MSc thesis, Hydroinformatics and Knowledge Management Program,UNESCO-IHE, Delft The Netherlands.

Abdullah, R. Price, Z. Vojinovic (2007), Spatial Decision Support System, "Flash Flooding and Knowledge Management", Water Knowledge Conference 2006, Belgium

Abdullah, R. Price, Z. Vojinovic (2007), Decision Support and Knowledge System for Urban Water Management "Case Studies in Flash Flooding and Receiving Water Impact", PhD Conference IHE 07, The Netherlands

Abdullah, M.Z. Abd. Rahman, R. Price, Z. Vojinovic (2007), Coupling Of GIS And DSS Into An Interactive Stormwater Management Information System, ISG 2007, Malaysia

Abdullah, R. Price, Z. Vojinovic (2007), Knowledge Management System "GIS Based Visioning, Strategizing and Scenario Module", European SWITCH Committee 2007, Swistzerland

Abdullah, A.A Rahman, Z. Vojinovic (2008), LiDAR Filtering Algorithms For Urban Flood Application, GISPRI 08, Malaysia

Abdullah, A.A Rahman, Z. Vojinovic (2009), LiDAR Filtering Algorithms For Urban Flood Application: Filters Test, Flood Modelling and Development of Filtering Algorithm, PhD Conference IHE 09, The Netherlands

Abdullah, A.A Rahman, Z. Vojinovic (2009), LiDAR Filtering Algorithms For Urban Flood Application: Filters Test, Flood Modelling and Development of Filtering Algorithm, 8UDM Conference 09, Tokyo

Abdullah, A.A Rahman, Z. Vojinovic (2009), LiDAR Filtering Algorithms For Urban Flood Application: Review On Current Algorithms, Filters Test and Urban Flood Modelling, Laserscanning Conference 09, France

Abdullah, A.F., Vojinovic, Z., Price, R.K., Aziz, N.A.A , (2011), A Methodology for Processing Raw LiDAR Data to Support Urban Flood Modelling Framework, Journal of Hydroinformatics, IWA Publishing 2011, doi:10.2166/hydro.2011.089.

Abdullah, A.F., Vojinovic, Z., Price, R.K., Aziz, N.A.A , (2011), Improved methodology for processing raw LiDAR data to support urban flood modelling: accounting for elevated roads and bridges, Journal of Hydroinformatics, IWA Publishing 2011, doi:10.2166/hydro.2011.009.

Acknowledgement

It is a pleasure to thank all those who have made this thesis possible. The research was funded by UNESCO-IHE under the SWITCH project and the Government of Malaysia. The data for this research was kindly provided by the Department of Irrigation and Drainage Malaysia (DID). The Mike Flood software for processing the hydrological and hydraulic data was kindly provided by UNESCO-IHE and Dr Nik and Associates Sdn. Bhd.

I would like to express my sincere gratitude to my advisor, Professor Roland Price for the ideas he had suggested for exploration, the discussions on the content, the sharing of thoughts on the philosophical aspects and the continuing efforts in correcting my English texts. I owe my deepest gratitude to my supervisor, Dr Zoran Vojinovic with whom I am working with not only on the current research but also on many other project activities within UNESCO-IHE. I am grateful for his enthusiasm, his inspiration and his great efforts in explaining things clearly and simply. Throughout my thesis writing period, he provided encouragement, sound advice, good teaching, good company and a lot of good ideas. I would have been lost without him. I am indebted to many of my colleagues who had supported and encouraged me throughout this research. I would also like to thank the members of the doctoral examination committee for evaluating this thesis.

Finally, I wish to thank my entire family for providing a loving and conducive environment for me. My wife and children especially were particularly supportive. Most importantly, I wish to thank my parents: they bore me, raised me, supported me, taught me and loved me. To them I dedicate this thesis.

Summary

In the last few decades, the consequences of floods and flash floods in many parts of the world have been devastating, with extensive tangible damages and unprecedented losses, personal pain, and social disruption. One way of improving flood management practice is to invest in data collection and modelling activities which enable an understanding of the functioning of a system and the selection of optimal mitigation measures. In this respect, the application of hydroinformatics technologies to urban water systems plays a vital role in making the best use of the latest data acquisition and handling techniques coupled with sophisticated modelling tools, including uncertainty analysis and optimisation facilities, to provide support to stakeholders for decision making. These technologies have revolutionized the way in which communication of information is carried out, with large amounts of data and information stored at nodes (servers) and accessible to anybody with a computer or mobile phone connected to the Internet from anywhere in the world.

Perhaps, out of all the data required by flood managers, a Digital Terrain Model (DTM) provides the most essential information. A DTM refers to a topographical map which contains terrain elevations, and as such is used to represent the terrain (or land surface) and its properties. Such DTM is a representation of the Earth's surface (or subset of it) and should strictly exclude features such as vegetation, buildings, bridges, etc. In urban flood management, DTMs are required for the analysis of the terrain topography and for setting-up 2D hydraulic models. Along with advances in computer power, researchers and practitioners have adopted more advanced modelling techniques, such as 1D/2D model coupling. This technique can be used to describe the dynamics and interaction between surface and sub-surface systems. For an efficient use of 2D models, the collection and processing of terrain data is of vital importance. Typically, Light Detection and Ranging (LiDAR) surveys enable the capture of spot heights at a spacing of 0.5m to 5m with a horizontal accuracy of 0.3m and a vertical accuracy of 0.15m. Most of LiDAR surveys result in a substantial amount of data, which requires careful processing before it can be used for any application. Recently, the vertical accuracy of LiDAR has increased dramatically to 0.05m (see for example FLIMAP, 2010). Filtering is a process of automatic detection and interpretation of bare earth and objects from the point cloud of LiDAR data, which results in the generation of a DTM. To date, many filtering algorithms have been developed, but none can yet be considered suitable to support reliable urban flood modelling work.

An assessment has been carried out to study the performance of seven different LiDAR filtering algorithms and to evaluate their suitability for urban flood modelling applications. It was found that none of these algorithms can be regarded as fully suitable to support such work in its present form. The research presents the augmentation of an existing Progressive Morphological filtering algorithm for processing raw LiDAR data to support a 1D/2D urban flood modelling framework. The key characteristics of this improved algorithm are: (1) the ability to deal with different kinds of buildings; (2) the ability to detect elevated road/rail lines and represent them in accordance to the reality; (3) the ability to deal with bridges and riverbanks; and (4) the ability to recover curbs and (5) the use of appropriated roughness coefficient of Manning's value to represent close-to-earth vegetation (e.g. grass and small bush).

The results of the improved algorithm were demonstrated using Kuala Lumpur (Malaysia) as a case study. The results have shown that the improved algorithm has better capabilities in identifying some of the features that are vital for urban flood modelling applications than any of the currently available algorithms and that it leads to better agreement between simulated and observed flood depths and flood extents. The overall results suggest that incorporation of building basements within the DTM, and that complete removal of elevated roads and bridges within the DTM, can lead to a difference in model results, which can, in some cases, be significant, with a tendency towards overestimating flood depth by those models which do not incorporate such a feature and when some other features are not properly removed. It is also suggested that the recovering of curbs within the DTM and the representation of close-to-earth by appropriate Manning's value can lead to some difference in model results, which may, in some cases, be significant with a tendency towards incorrect flood flow by those models in which such features are not properly represented.

Samenvatting

De afgelopen decennia zijn de gevolgen van hoge waterstanden en plotselinge overstromingen in vele delen van de wereld verwoestend geweest, met enorme materiële schade en ongekende verliezen, persoonlijk leed en maatschappelijke ontwrichting tot gevolg. Een manier om overstromingen te beheersen en te reduceren is te investeren in het verzamelen van gegevens en het opzetten van modelsystemen teneinde inzicht te krijgen in het functioneren van het systeem en de selectie van optimale risicobeperkende maatregelen mogelijk maken.De toepassing van hydroinformatics technieken op stedelijke watersystemen speelt hierbij een vitale rol met name voor het verkrijgen van de meest recente data inclusief geavanceerde verwerkingstechnieken , waaronder onzekerheidsanalyse en optimalisatietechnieken, ter ondersteuning van belanghebbenden bij besluitvorming. Deze technologieën hebben een ommekeer teweeggebracht in de manier waarop de communicatie van informatie wordt uitgevoerd, met grote hoeveelheden gegevens en informatie die is opgeslagen in databestanden op computerservers en die toegankelijk zijn voor iedereen ter wereld met een op het internet aangesloten computer of mobiele telefoon.

Misschien biedt een Digitaal Terrein Model (DTM) van alle vereiste gegevens voor overstromingsbeheerders wel de meest essentiële informatie. Een DTM verwijst naar een topografische kaart die terreinhoogtes bevat, en als zodanig wordt gebruikt om het land oppervlak met bijbehorende eigenschappenaf te beelden. Een dergelijk DTM is een weergave van het aardoppervlak (of een gedeelte daarvan) en zou strikt genomen kenmerken zoals vegetatie, gebouwen, bruggen etc. moeten uitsluiten. DTM's zijn vereist voor de analyse van de topografie van het terrein en voor het opzetten van 2D hydraulische modellen voor stedelijk waterbeheer. Vanwege de vooruitgang in rekenkracht, hebben onderzoekers en vakspecialisten steeds geavanceerdere modelleringtechnieken, zoals 1D/2D model koppeling, toegepast. Deze techniek kan worden gebruikt om de dynamiek en interactie tussen het oppervlak en de ondergrond te beschrijven. Om efficiënt gebruik te maken van 2D modellen is de verzameling en verwerking van terreingegevens van vitaal belang. Kenmerkend voor Licht Detectie en Rangschikking (LiDAR) metingen is het herkennen van de hoogte ter plaatse op een afstand van 0,5 m tot 5 m met een horizontale nauwkeurigheid van 0,3 m en een verticale nauwkeurigheid van 0,15 m. De meeste resultaten van de LiDAR metingen resulteren in een aanzienlijke hoeveelheid gegeven, die zorgvuldig verwerkt moeten worden voordat ze kunnen worden toegepast. Onlangs is de verticale nauwkeurigheid van LiDAR dramatisch toegenomen tot 0,05 m (zie bijvoorbeeld FLIMAP, 2010). Filteren is een proces van automatische detectie en interpretatie van kale bodems en objecten uit de puntenwolk van LiDAR gegevens, die in het genereren van een DTM resulteert. Tot op heden hebben zijn veel filtering algoritmes ontwikkeld, maar geen van deze kan nog worden beschouwd als geschikt ter ondersteuning van betrouwbare stedelijke overstromingsmodellen. In dit proefschrift zijn de prestaties van zeven verschillende LiDAR filter algoritmes bestudeeerd en op hun geschiktheid beoordeld voor toepassing in stedelijke overstromingsmodellen. Het bleek dat geen van deze algoritmen in zijn huidige vorm kan worden beschouwd als volledig geschikt voor dit doel. Het onderzoek laat een uitbreiding zien van een bestaand Progressief Morfologisch filtering algoritme voor de verwerking van onuitgewerkte LiDAR gegevens om 1D/2D stedelijke overstromingsmodellering te ondersteunen.

De belangrijkste kenmerken van dit verbeterde algoritme zijn: (1) de mogelijkheid om rekening te houden met verschillende soorten gebouwen; (2) de mogelijkheid om verhoogde weg/spoor lijnen op te sporen en weer te geven in overeenstemming met de werkelijkheid; (3) de mogelijkheid om bruggen en rivieroevers in de schematisatie op te nemen; en (4) de mogelijkheid om rekening te houden met verhoogde drempels en wegen, alsmede (5) het bepalen van geschikte ruwheids coëfficiënten (Manning's waarden) van 'lage' vegetatie (zoals gras en struiken).De resultaten van het verbeterde algoritme worden gedemonstreerd aan de hand van een toepassing in Kuala Lumpur (Maleisië). De resultaten tonen aan dat het verbeterde algoritme beter in staat is om die functies te identificeren welke van vitaal belang zijn voor de toepassing in stedelijke overstromingsmodellen dan andere momenteel beschikbare algoritmen, hetgeen leidt tot betere overeenkomst tussen berekende en waargenomen overstromingsdiepten en de omvang van de overstromingen. De algemene resultaten suggereren dat het niet meenemen van kelders van gebouwen in de schematisatie, en het volledig verwijderen van verhoogde wegen en bruggen in de DTM, kunnen leiden tot aanzienlijke verschillen in resultaten van het model, met een neiging tot het overschatten van de overstromingsdiepte door modellen die dergelijke functies niet in zich hebben of wanneer die functies niet correct zijn meegenomen.

Er wordt ook gesuggereerd om drempels in de DTM mee te nemen en de aanwezigheid van 'lage' vegetatie in de Manning waarde op te nemen, aangezien dit kan leiden tot – in sommoge gevallen aanzienlijke – verschillen in modelresultaten, met als gevolg onjuiste stroomafvoer door modellen waarin dergelijke functies niet voldoende aanwezig zijn.

Table of Content

ix

List of Figures

List of Tables

Chapter 1

Introduction

1.1 Background

In the last few decades, the consequences of floods and flash floods in many parts of the world have been devastating, with extensive tangible damages and unprecedented losses, personal pain, and social disruption. Floods can be described according to their speed (such as flash or lowland floods), geography (rural or urban), or their cause (precipitation or dam-break). The different types of flooding include flash floods, coastal floods, urban floods, river (fluvial) floods and ponding (pluvial) floods. One way of improving flood management practice is to invest in data collection and modelling activities, which enable an understanding of the functioning of a system and the selection of optimal mitigation measures. In this respect, the application of hydroinformatics technologies plays a vital role in making the best use of the latest data acquisition and handling techniques, coupled with sophisticated modeling tools, including uncertainty analysis and optimization facilities to provide support to stakeholders for decision making (Price & Vojinovic, 2010). These technologies have revolutionized the way in which the communication of information is carried out, with large amounts of data and information stored at nodes (servers) and accessible to anybody with a computer or mobile phone connected to the Internet anywhere in the world (see also Abbot & Vojinovic, 2009).

Within the flood management process, data acquisition refers to the compilation of existing data and the collection of additional data for system analysis, modelling and decision making. A typical flood management database consists of spatial, temporal and other data (e.g. design standards, flood incidents, public perception of a utility's levels of service, etc.). The collection of such data is of the utmost importance for making cost-effective investment and operational or maintenance decisions. The role of modelling within urban flood management is in complementing the acquisition of data to improve the information and understanding about the performance of a given drainage network, taking into account the associated urban terrain. Considerable attention has been given to the acquisition of good geometric and topographical data at adequate resolution in order to describe the primary features of the flow paths through the urban area. In this respect, a Digital Terrain Model (DTM) represents one of the most essential sources of information that is needed by flood managers. A DTM refers to a topographical map, which contains terrain elevations, and as such, it is used to characterise the terrain (or land surface) and its properties. It is a representation of the Earth's surface (or subset of it) and it strictly excludes features such as vegetation, buildings, bridges, etc. In urban flood modelling,

DTMs are required for the analysis of the terrain topography, identification of overland flow paths, setting-up 2D hydraulic models, processing model results, delineation of flood hazards, producing flood maps, estimating damages, and evaluating various mitigation measures. Nowadays, one of the most preferred techniques for modelling floods in urban areas is a coupled 1D/2D modelling approach. This technique can be used to describe the dynamics and interaction between surface and sub-surface systems. For an efficient use of 2D models the collection and processing of terrain data is of vital importance.

Typically, a DTM data set can be obtained from ground surveys (e.g. total stations together with Global Positioning System – GPS), aerial stereo photography, satellite stereo imagery, airborne laser scanning or by digitizing points/contours from an analogue format such as a paper map, so that they can be stored and displayed within a GIS package and then interpolated. Airborne laser scanning (ALS) or light detection and ranging (LiDAR) is one of the most common techniques that is used to measure the elevation of an area accurately and economically in the context of cost/benefit analysis. Such an analysis compares the costs of providing facilities to reduce the frequency or degree of flooding against the damage that would be incurred. The airborne measurement devices can deliver information on terrain levels to a desired resolution. The end result of a LiDAR survey is a large number of spot elevations which need careful processing. Typically, thinning, filtering and interpolation are techniques that need to be adopted as part of this process. Thinning (or reduction of data points) is usually achieved by removing neighbouring points that are found to be within a specified elevation tolerance. Filtering is a process of automatic detection and interpretation of bare earth and objects from the point cloud of LiDAR data, which results in the generation of a DTM. To date, many filtering algorithms have been developed, and in a more general sense, many of them have become standard industry practice. However, when it comes to the use of a DTM for urban flood modelling applications, these algorithms cannot be always considered suitable. Depending on the terrain characteristics, they can even lead to misleading results and degrade the predictive capability of the modelling technique. This is largely due to the fact that urban environments often contain a variety of features (or objects), which have the ability to store or divert flows during flood events.

As these objects dominate urban surfaces, appropriate filtering methods need to be applied in order to identify such objects and to represent them correctly within a DTM so that the DTM can be used more safely in modelling applications. However, most of the current filtering algorithms are designed to detect vegetation and freestanding buildings only, and features such as roads, curbs, elevated roads, bridges, rivers and river banks are always difficult to detect. Therefore, further improvements of LiDAR filtering techniques are needed so that these features can bed detected and modelling can generate more fruitful results. In this research, the filtering methods and their application are focused on urban areas, and consequently on aspects of urban flooding.

1.1.1 Urban floods

The rapid expansion of urban areas over the past two decades has resulted in the extensive growth of many population centres and settlements along coastal and low-lying areas that are prone to flooding. As a consequence, the frequency and severity of flooding have increased at both basin and local levels; particularly in these urban areas. Significant flood losses have frequently been experienced in many major urban centres, such as Bangkok, Dhaka, Hanoi, Jakarta, Kuala Lumpur, Manila, and Phnom Penh. Flooding in urban drainage systems may occur at different stages of hydraulic surcharge, depending on the type of drainage system (separate or combined sewers), their general drainage design characteristics, as well as specific local constraints. Rapid developments in urban areas change the land coverage from vegetation to buildings and roads, which lead to a related change in the hydrologic system of the basin. Impervious concrete and asphalt surfaces that cover most of the urban areas decrease the ability of the land to absorb rainfall, and force the excess rainfall-runoff to flow faster over the surface. Blockages within a drainage system, inadequate capacity of drains, and heavy precipitation, can be the main causes of urban flooding. However, in many major urban centres, flooding may also be caused by excess flows in streams flowing through the urban area, which originate from rural catchments. Such flooding can be exacerbated by conjunction with riverine flooding. Whilst riverine flooding usually requires large-scale measures, local flooding caused by urban streams offers opportunities for small-scale measures to be adopted, and for the local communities to play an active role in flood management. Such flooding may still cause great damage to residential and commercial buildings or other public and private infrastructures. These flood damages are expected to increase in the future with the continued urban expansion and escalation of land and property values.

1.1.2 Effect of urban flooding

Urban flooding causes considerable damage and disruption, with serious social and economic impacts. The effects of flooding vary, due to local physical, geographical, and meteorological conditions, and therefore, each situation requires an individual response (Smith, 2009). Direct flood damage covers all varieties of loss to individuals and communities relating to the immediate physical contact of floodwater to human beings, property, and the environment. This includes damage to buildings, property, dikes, destruction of standing agricultural crops, loss of livestock and human life, immediate health impacts, and the contamination of ecological systems. Indirect or consequential effects of flood damage occur as a further consequence of the flood and the disruption of economic and social activities. This damage can affect areas larger than those actually inundated.

One prominent example of indirect effects is the loss of economic production due to destroyed facilities, breakdown in energy, and telecommunication supplies, and the interruption of supply chains. Other examples include loss of business due to traffic disruption, disturbance of markets (e.g. higher prices for food or decreased prices for real estate near floodplains), reduced productivity together with the consequence of decreased competitiveness in selected economic sectors or regions, and the disadvantages connected with reduced markets and public services

Konig et al., (2002), divided urban flooding damages into the following categories:

i. Direct damage: typically material damage caused by water or flowing water
ii. Indirect damage: e.g., traffic disruption, administrative and labour costs, production losses, spreading of disease, etc.
iii. Social consequences: negative long-term effects of a more psychological character, such as the decrease of property values in frequently flooded areas and delayed economic development.

The extent of economic loss depends mainly on indirect factors determined by both the flood and the area characteristics. Direct damage consists of damage to property, services and production. These direct losses also include the loss of production for a certain period of time after the flooding. In order to minimize the damage caused by urban flooding, efficient urban flood management is essential. A reduction in the damage costs (regarded as a benefit) through structural or non structural measures (with their corresponding cost) forms the basis of a cost-benefit analysis in which the objective is to minimise the sum of the cost and the corresponding benefit.

1.1.3 Urban flood management

Within flood research, it has been widely accepted that absolute flood protection cannot be achieved (Schanze, 2006). Instead, growing attention has been given to a new paradigm of flood management, based on the effective establishment of both risk mitigation (structural, technical flood defence measures, such as dams, dikes, or polders) and adaptation (non-structural, 'soft' measures, such as preparation of the local people, flood insurances, information management, and social networks) measures (Kubal et al., 2009).

Parkinson J and Mark O (2005), in their book summarise the short, medium, and long-term objectives, of storm water management strategies. In the short-term, priorities include runoff control flood protection and pollution mitigation strategies, which in many developing countries have yet to be addressed effectively. The medium-term objective focuses on development and implementation of water quality improvement, water conservation, and a strategy to preserve the hydrology of the natural catchment. The long-term objective places a greater emphasis on the preservation of natural resources, the amenity value of water in an urban environment for recreational activities, and the promotion of an increased awareness of environmental issues.

Although these objectives may initially appear to be somewhat idealistic goals, especially considering the existing situation in many developing countries, it is important that planners and designers of urban drainage systems aim to satisfy the needs of future generations by adhering to the objectives of sustainable development, as defined by the World Commission on Environment and Development, in 1987.

The flood management process can be divided into three phases (Ahmad S and Simonovic S.P., 2006, Vojinovic, and van Teffeelen, 2007):

    i.   Pre-flood preparation and planning - Different flood management options, including structural and non-structural, are analysed and compared for possible implementation to reduce flood damage.

    ii.   Flood emergency management - Involves the forecasting of floods and a regular updating of forecasts.

    iii.   Post-flood recovery - Involves decisions regarding the return to normal life and activities after a period of flooding.

Structural flood mitigation works are usually expensive, and create social disruption and inconvenience during the construction period. Hence, in most cases, the optimal strategy for flood control is one which combines structural measures with non-structural measures, developed on the basis of a comprehensive master plan study that takes into account the future potential for development and land use. There needs to be a wider application of planning and legal instruments through appropriate laws and administrative procedures as these would ensure that future development takes place with the least burden or impact on existing drainage systems, particularly in highly built-up areas where land acquisition, construction, and utility reallocation costs are high or prohibitive. The problem of urban drainage and flooding is of real concern and must be dealt with properly since it directly impacts quality of life and the living environment, as well as supporting and sustaining urban growth.

A non-structural flood mitigation strategy relies upon action and support from households and local organizations working collaboratively, and requires the participation of inhabitants of areas prone to flooding. In addition, flood warning systems need to be in place so that warnings can be issued to prepare communities for the onset of a large flood event and for the urban authorities to be on alert during an emergency situation. Although these response strategies can minimize potential damage, there will also be a need to develop appropriate strategies for flood recovery and rehabilitation of the affected communities.

Neo Tong Lee (1995) states that as a country develops, flood mitigation works become larger (in terms of human and financial resources), and it becomes necessary to address the problem seriously. In particular, a wide range of technical and economic options/instruments need to be studied for the following reasons:

i.   Engineering options are increasingly expensive, particularly when the improvement works have to be constructed in highly built-up areas. In addition, there are practical difficulties in relocating squatters, services, and public utilities, as well as the limitations of working space. All of these factors contribute to the spiralling cost of structural measures for flood control.

ii.   In many urban centres, the cost of acquiring land or reserves for the construction of drains has become a major concern. There have been situations where the cost of land acquisition was higher than the engineering construction cost.

iii. State governments, local authorities, and the private sector should assume a larger role and responsibility in addressing storm water management problems. For example, there is a general reluctance among housing developers and local authorities to set aside flood retention areas, because this will reduce the net area available for urban development. As a result, the peak flow in a main drain or river system keeps increasing, as the upper catchment areas are progressively developed or urbanized.

iv.   There is a need to promote the concept and practice of cost recovery from direct beneficiaries of drainage improvement works. At present, only six states in Malaysia (Selangor, Melaka, Negeri Sembilan, Johor, Penang, and Kedah) collect drainage contributions from housing developers. Even then, the current rates are grossly inadequate to cover the actual cost of urban drainage improvement works.

It is commonly accepted that proper drainage of stormwater and protection against flood losses are fundamental for the sustained development and growth of modern cities. However, there are technical and economic constraints on the provision of structural measures to control urban flooding. Sustainable management of urban stormwater involves water conservation, pollution prevention, and ecological restoration. Goals for sustainable management include flood reduction, pollution minimization, stormwater retention, urban landscape improvement, and the reduction of drainage investments. Such a reduction can be achieved using various methods, such as minimizing peak stormwater discharges from urban catchments, managing pollution loads, harvesting rain and stormwater runoff, functionally incorporating stormwater into urban streetscapes, promoting green areas and innovatively integrating stormwater systems into the urban environment, thereby reducing the cost of infrastructure (Brown, 2005). For the successful implementation of these measures, it is necessary to ensure that all stakeholders fully understand the causes of urban flooding and recognize that the financial and environmental implications are based on sustainable economic development and sound environmental management. In order to understand and manage floods better in urban areas, it is important to be able to reproduce the flood physics, which consists of the flow over a surface area (i.e. the floodplain) and the flow in drainage systems, which are often below the ground. Understanding the drainage system's function gives advantages in evaluating alleviation schemes and choosing the optimal scheme, which is to be implemented to solve the flooding problem. The most common and acceptable way of providing this information is by using urban flood modelling.

1.1.4 Urban flood modelling

Typically, urban flood modelling practice concerns the use of 1D, 1D/1D, 2D and 1D/2D modelling approaches; see for example, Chen, et al., (2005), Garcia-Navarro and Brufau (2006), Hunter et. al. (2007), Hunter et al. (2008), Kuiry et. al. (2010) and Price and Vojinovic (2011). Mark et al. (2004) demonstrated how a 1D modelling approach can be used to incorporate the interaction between (i) the buried pipe system, (ii) the streets (with open channel flow) and (iii) the areas flooded with stagnant water. Djordjevic, et al., (2005) have implemented a dual drainage concept (which consists of a combination of minor and major systems) within a 1D model. Vojinovic and Tutulic (2009) have explored the difference in predictive capabilities of 1D and 1D/2D modelling approaches for the purpose of urban flood analysis across irregular terrains and their corresponding damage estimation. Also, Vojinovic et. al (2011) have shown how different terrain data resolution, features such as roads and building structures, and different friction coefficients can affect the simulation results of 1D/2D models. The literature to date confirms that apart from different model formulations, the variations in the ground topography, discontinuities, representation of features, surface roughness and terrain data resolution are important factors that need to be carefully considered and accounted for in flood modelling studies.

Typically, 1D models are used to simulate flow through pipes, channels, culverts and other defined geometries. The system of 1D cross-sectionally-averaged Saint-Venant equations, which are used to describe the evolution of the water depth and either the discharge or the mean flow velocity, represent the principles of conservation of mass (continuity equation) and momentum. The boundary conditions are the discharges, water levels (or depths) or free flow conditions at the ends of the conduits or channels. In a channel network, the boundary values of the dependent variables such as the discharge or the depth are not known in advance and need to be determined by a solution procedure for the numerical form of the analytical equations. The solution is commonly based on a temporary elimination of variables at internal cross-sections and the reduction of the numerical equations to a system of unknown water levels at the junction nodes of the network.

The system of 2D shallow water equations consists of three equations: one (1) continuity, and two (2) equations for the conservation of momentum in two Cartesian coordinates. The simulation process in the case of coupled 1D-2D modelling is based on complex numerical solution schemes for the computation of water levels, discharges and velocities. The surface model (i.e., 2D model) simulates vertically-integrated two-dimensional unsteady flow given the relevant boundary conditions and calibration parameters (e.g. resistance coefficients, etc.) and the bathymetry (as provided by a digital terrain model of the catchment area). The interactions between channels and floodplains are determined according to the type of link between them. For example, discharges generated by pumping stations, weirs or orifices are regarded as lateral inflows to the 2D model. Also, if the channel or conduit flows exceed the ground level (for pipe network systems) or bank levels (for open channels) then the discharge is computed by a weir (or orifice) discharge equation and it is also considered as a lateral inflow to the 2D model.

The two domains (1D and 2D) are normally coupled at grid cells overlying the channel computational points through mutual points of the connected cell and the adjoining channel section (Vojinovic and Tutulic, 2009; Price and Vojinovic, 2011). The dynamic link that allows for interaction of mass and momentum fluxes between the two model domains has been implemented in several commercial software packages. Such examples are MIKE11 for the 1D modelling system and MIKE 21 for the 2D modelling. Solving for water flows on a regular grid, as in the case of MIKE 21, has the advantage of providing an easy integration with DTMs which are most often available in a regular grid format. In this research, the modelling part is carried out by the 1D/2D modelling technique using the above mentioned commercial packages: MIKE11 and MIKE21.

1.1.5 Light Detection and Ranging (LiDAR)

LiDAR is an optical remote sensing technology that measures the properties of scattered light in order to find the range and/or other information of a distant target. The prevalent method to determine the distance to an object or surface is to use laser pulses. Similar to radar technology, which uses radio waves, the range to an object is determined by measuring the time delay between the transmission of a pulse and detection of the reflected signal. LiDAR is a surveying method which can give accurate x, y, and z (height) positions. There are several techniques of obtaining height remotely, such as aerial photos and Interferometric Synthetic Aperture Radar (InSAR). However, LiDAR can offer the accurate positioning of large areas, which is both cost and time effective.

LiDAR collects highly accurate and a dense set of points from the surface or terrain. This data is collected using aircraft-mounted lasers capable of recording elevation measurements at a rate of 2,000 to 5,000 pulses per second and which have a vertical precision of 15 centimetres. Collected sample points represent the height of a terrain. The points collected from LiDAR are highly accurate and dense, which enables the construction of a Digital Terrain Model (DTM). In order to construct the DTM from raw data, particular pre and post-processes need to be performed, including the modelling of systematic errors, filtering, and thinning. Of these data processes, filtering poses the greatest challenge.

1.1.6 LiDAR filtering algorithm

Filtering during the LiDAR process can be defined as distinguishing and separating objects and ground surface points. Recently, a number of algorithms have been developed to enhance the extraction process in which surface points are separated from the point clouds obtained by LiDAR. Most of these filtering algorithms classify data into at least two categories, namely object and surface, using local neighbourhood operations on points (Sithole and Vosselman, 2004). Every filter makes an assumption about the structure of the bare earth points in a local neighbourhood. For example, bare earth points in a locale must fit a given parametric surface. Four distinct groups of filters can be identified:

The first group is the so called, mathematical morphological filtering. The second group can be described as progressive filters. In the third group of algorithms, a Triangulated Irregular Network (TIN) is derived from neighborhood minima, and then the laser point cloud is progressively made more dense. This process is called densification. The fourth group of filters consists of segmentation-based filters where every pixel is classified into ground or not-ground, based on segmentation.

The main distinction between these different approaches can be seen in the strategy that they use to estimate the planimetric and height differences between object and surface points. Additionally, until now, most of these algorithms have been built for a rural environment and very few have been focused on separating urban objects from the complex urban surface. The filtering process for an urban environment is essential in order to produce an accurate DTM, especially when the DTM is used as input to an urban flood model.

1.2   Research methodology

This research begins with a literature review, which covers important subjects, such as urban flood modelling, LiDAR, and filtering algorithms. The implementation of the research, including methodology, filtering algorithm, and software, are based on literature. Data was collected from various sources, such as the Department of Irrigation and Drainage Malaysia, Survey Department of Malaysia, and private companies. The data is divided into two groups. The first is topographical data, such as DTM, aerial photo or satellite images, cadastral map, street map, location of rainfall stations, catchments, land use, building databases and survey control points. The second group of data covers hydrology and hydraulics, including rainfall time series, water levels, channels, drainage, rivers, pipes, pumping stations, and detention ponds.

The use of an urban flood model is essential in order to test the accuracy of the developed LiDAR filtering algorithm. Several 1D-2D urban flood models are developed using the DTM, which is processed by different algorithms. The first experiment uses the best existing data available, such as LiDAR data; the second uses an existing filtering algorithm, and the final experiment uses a newly developed algorithm. The urban flood model with the DTM was developed using existing commercial software from DHI Water & Environment, such as MIKE11, MIKE 21, and MIKEFlood.

The new algorithm is developed based on urban characteristics. The major task of this algorithm development is to define the rules that distinguish surfaces and objects in an urban area. In order to do this, current algorithms are studied and their weaknesses of current assumptions are taken into account. Defining the rules is difficult due to the complexity of an urban area landscape. Arising from insights into the current assumptions, new assumptions are formulated which are more suitable for an urban environment. Classification of the different types of objects is done later; related to the contents of the landscape, such as buildings, roads, and bridges. Such a classification is performed at several levels until valid objects and surfaces are detected and determined. For the modified algorithm, the detection and determination of some objects and surfaces are improved using data fusion. With data fusion, the point cloud is overlaid with an aerial photo and existing data so that the detection of objects and surfaces can be done quickly. The improvements are focused on the following issues:

i.   To detect buildings and to classify them into: solid buildings, passage buildings and buildings with basements;
ii.   To detect elevated road/rail lines, to remove them from a DTM and to incorporate any structures underneath them;
iii. To detect and cross-reference the location of bridges for the purpose of setting up a 1D model and to remove the bridges from the DTM that will be used to set up the 2D model;
iv.   To apply a data fusion concept and combine the river polygon data with the LiDAR data for the purpose of identifying riverbanks and interpolating between the points along the river banks when the river network is modelled with a 1D model such as MIKE11;
v.   To retain curbs and recover the discontinuities introduced by the curbs
vi.   To detect and remove tall vegetation with a height of more than 0.3m;
vii. To assign a roughness coefficient to an area containing closed-to-earth vegetation e.g. grass and shrub with a height of less than 0.3m;
viii. To generate a DTM for use with a 2D flood model such as MIKE21.
ix.   To test the usefulness of the algorithm by carrying out 1D/2D modelling of flooding for a study area.

The results of the filter are analysed in two ways: by qualitative and quantitative assessments. Qualitative assessment is based on the performance of the algorithms in surmounting several difficulties. The performances of the algorithms are based on a visual examination and comparison between the filtered datasets in order to see which filter algorithm is likely to fail. The quantitative assessment is based on the evaluation of errors and the analysis of the terrain data. This assessment leads to the generation of tables to evaluate errors and to determine the relationship between the errors.

Calibration of the urban flood model is performed to ensure that the model can produce reliable results. The calibration of simulation models must be done carefully due to the simplistic nature of the model components used to represent complex physical processes in urban areas. Calibration is performed by comparing the model results with historical events. A good agreement between simulated and observed hydrographs increases confidence in the model performance - at least for events similar in magnitude to those simulated. Calibration is also performed by comparing the simulated flood extent with the recorded flood extent from the same area. The calibration and verification of the urban flood model is performed using the precipitation and event datasets archived by the Malaysian Department of Irrigation and Drainage (DID).

1.3   Research objectives

The overall aim of this research is to develop a suitable LiDAR filtering algorithm for urban flood modelling. This includes the development of a 1D-2D flood simulation model, the identification of the uncertainties in the simulation, a review of existing LiDAR filtering algorithms, the production of a new LiDAR filtering algorithm that best suits the purpose of urban flood modelling, and the development of a graphical user interface. The focus is on understanding current filtering algorithms and modifying them to produce the best DTM that can closely replicate the urban topography.

The more specific objectives are:

i.   To review existing LiDAR filtering algorithms and evaluate their advantages and disadvantages
ii.   To identify the preferred algorithm for improvement
iii.   To set up 1D-2D flood simulation models for an area case study
iv.   To improve the preferred algorithm and verify its performance through a case study:
v.   detect buildings and to classify them into three categories
vi.   detect and remove elevated roads and incorporate ground structures underneath them;
vii.   detect and cross-reference the location of bridges and remove them from the DTM to facilitate a 1D simulation model;
viii. combine the river polygon data with the LiDAR data for the purpose of identifying riverbanks and interpolating the points between them;
ix.   retain curbs and recover the discontinuities in the ground surface introduced by the curbs;
x.   detect and remove tall vegetation (height of more than 0.3m) and assign a roughness coefficient to the area containing close-to-earth vegetation (height less than 0.3m).

1.4   Contribution to knowledge

This research into the development of a new filtering algorithm contributes to the field of Airborne Laser Scanning and urban flooding in the following ways:

i.   Development of a new filtering algorithm designed specifically for urban flood modelling.

ii. The deduction of a novel approach of attaching depth cavities for buildings with basements. The standard approach for dealing with buildings is either to completely remove them from the DTM or to rebuild them as solid objects. The approach taken here is to attach the building with basement with the basement properties. This is achieved by lowering the area of the building to a specific height below ground level. The determination of this height depends significantly on the structure and the environment of the study area. This condition produces the so-called 'retention pond imitation' in which flooding can inundate the building area first before it floods the surrounding area.

iii. Detection of elevated roads and train lines using a combination of slope and signal intensity. Most of the current filtering algorithms do not distinguish between roads and elevated roads. It is essential to detect and remove elevated roads from the DTM because their long linear form acts as a wall which diverts the flood flow, even though in real situations the water can pass through. This research uses the concept of 'slight' and 'hi' to detect high objects. Elevated roads are separated from other objects using the intensity value of asphalt (the material used in road construction).

iv. Detection of curbs and the recovery of a curb's discontinuity. A 1m x 1m grid is used as an input, as the flood urban model cannot represent curbs perfectly, making the discontinuity the main problem. The new filtering algorithm uses the advantage of a higher resolution grid to recover the discontinuity of the curbs by adding them to the area such that the discontinuities occur as a sub-grid.

v. Assigning a roughness coefficient to the area containing close-to-earth vegetation. Not all vegetation is removed from the DTM by the filtering algorithm due to limitations of the algorithm especially for vegetation that lies close to the earth like grass and shrub. This area is detected and assigned with an appropriate roughness coefficient.

1.5 Outline of the thesis

This thesis has nine chapters. Chapter 2 discusses the methods and equations involved in urban flooding modelling. It covers both 1D and 1D/2D coupling models. Several existing flood models for urban applications are reviewed. Issues and discussions on urban flood modelling, such as data requirements, accuracy, robustness, and more, are also covered in this chapter.

Chapter 3 discusses the two dimensional (2D) surface model. A detailed explanation focuses on the 2D surface model, which is the most important input data for the 1D/2D coupling model. The 2D surface model represents the physical urban surface, particularly where the overflows from river and drainage floods the land. The Digital Surface Model and the Digital Elevation Model, are also discussed, including concepts, components, sources, and the differences between the models. This chapter explains the nature of the essential objects on the urban surfaces and their impact on the urban flood model.

Chapter 4 reviews the ALS. Starting with a definition, this chapter discusses in detail the principle of ALS and its components. It also discusses LiDAR Intensity, which as a side product is used later in this research. LiDAR processing includes the modelling of systematic errors, filtering and classification, feature detection, and thinning. These issues and discussions on ALS such as accuracy, uncertainty, and errors are also included in this chapter.

Chapter 5 discusses the concept of the filtering algorithm. Filtering and classification of bare earth and objects are discussed.

Chapter 6 reviews seven selected existing filter algorithms. This chapter discusses the weaknesses of current filters in terms of how they handle landscape and data characteristics. This is followed by a discussion on the weaknesses of current assumptions, which are the basis for formulating the new assumption suitable for urban flood modelling purposes. Each filter is evaluated using a quantitative and qualitative assessment, and the results are compared and explained. The similarities and differences, including the advantages and disadvantages of all filters, are discussed. Based on the review and the study, this chapter explains why the Progressive Morphological Algorithm has been selected to be modified in the formulation of the new filter.

Chapter 7 covers the formulation of the new filter by a modification of the Progressive Morphological Algorithm. The framework of the new filter is discussed, and the methods and approaches used are explained step-by-step. The modification process for each object is also discussed in a different section so that the method and approaches used can be explained clearly. Limitations, accuracy, and the robustness of the new filter are also discussed at the end of this chapter.

Chapter 8 discusses the case study performed in this research by covering the selected data, model structure, model set-up, and flood events. The comparisons between the results of the flood model using the DTM with the new filter, and the results of the flood model using the DTM with the current filter, are discussed in terms of flood depth, flood extent, and flood velocity. The overall results, including which model gives the closest reading to the recorded data, are discussed at the end of this chapter.

In Chapter 9, the conclusion from the overall research is presented. Recommendations for future work are also covered in this chapter.

Chapter 2

URBAN FLOOD MODELLING

2.1   Introduction

An urban flood model uses a numerical representation of the flow. A hydrodynamic model of flooding is used to simulate the flow and represent how the flood could happen. The hydrodynamic model can be identified by the spatial complexity of the schematization such as 1D, 2D and 3D, the grid types such as rectangular, curvilinear, and finite volumes or elements, and the numerical solution of the flow equations. Research shows that 1D/2D coupled models can represent urban flooding much better than using only a 1D model. This is because the simulation of the urban flooding mechanisms, which consist of the flow over a surface area, the flow in the drainage system and the flow in other underground structures, is essential to describe the real flooding process in an urban area. Additionally, the 1D flow technique is only valid under the assumption that flows are varied in the longitudinal direction but not when the flood occurs on the street where it also involves flow in the transverse direction. Because of this, it is seen that the flood modeling approach is leading towards a new era in which an urban flood model is now formed by the coupling model between the overland surface flow and possibly the urban stream network and the underground pipe flow.

2.2   Theoretical background on urban flood modeling

An urban flood model can normally be divided into two major sub-models which are inter-related. These sub-models include:
    i.   Hydrological model
    ii.   Floodplain model

2.2.1 Hydrological modeling

Hydrological models are simplified, conceptual representations of part of the hydrologic cycle. They are primarily used for hydrologic prediction and for understanding hydrologic processes. In the context of urban flood modeling, the main hydrological model used is for rainfall-runoff. The rainfall-runoff is the flow of excess water over the land surface that occurs when the soil is infiltrated to full capacity.

Rainfall-runoff models may be grouped into two general classifications, which are illustrated in Figure 2.1 and Figure 2.2. The first approach uses the concept of effective rainfall in which a loss model is assumed. This divides the rainfall intensity into 'losses' and an effective rainfall hyetograph. The effective rainfall is then used as the input to a catchment model to produce the runoff hydrograph. It follows from this approach that the infiltration process ceases at the end of the storm duration.

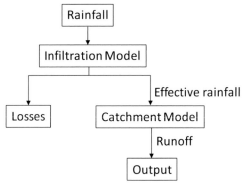

Figure 2.1: A rainfall-runoff models using effective rainfall

An alternative approach that might be termed a surface water budget approach incorporates the loss mechanism into the catchment model. In this way, the incident rainfall hyetograph is used as an input, and the estimation of infiltration and other losses is made as an integral part of the calculation of the runoff. This approach implies that infiltration will continue to occur as long as the average depth of excess water on the surface is finite. Clearly, this may continue after the cessation of the rainfall.

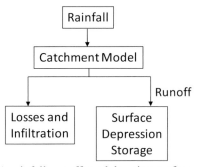

Figure 2.2: A rainfall-runoff models using surface water budget

2.2.3 Floodplain modeling

For floodplain modelling, most of the recent professional engineering software packages contain a comprehensive modelling system for 2D free surface flows. The models generated are applicable to the simulation of hydraulic and related phenomena in lakes, estuaries, bays, coastal areas and seas where stratification can be neglected.

Most of the models are based on a numerical finite difference solution of full non-linear equations of conservation of mass and momentum integrated over the vertical to describe the flow and water level variations. The application of an implicit finite difference scheme results in a tridiagonal system of equations for each grid line in the model. The solution is obtained by inverting the tridiagonal matrix using a Double Sweep algorithm, which is a very fast and accurate form of Gaussian elimination.

The following equations for the conservation of mass and momentum are integrated over the vertical to describe the flow and water level variations:

Continuity
$$\frac{\partial \zeta}{\partial t} + \frac{\partial p}{\partial x} + \frac{\partial q}{\partial y} = 0$$
(Equation 2.1)

x-momentum
$$\frac{\partial p}{\partial t} + \frac{\partial}{\partial x}\left(\frac{p^2}{h}\right) + \frac{\partial}{\partial y}\left(\frac{pq}{h}\right) + gh\frac{\partial \zeta}{\partial x} + \frac{gp\sqrt{p^2+g^2}}{C_2 h_2} -$$
$$\frac{1}{\rho_w}\left[\frac{\partial}{\partial x}(h\tau_{xx}) + \frac{\partial}{\partial y}(h\tau_{xy})\right] - \Omega q - fVV_x + \frac{h}{\rho_w}\frac{\partial}{\partial x}[p_a] = 0$$
(Equation 2.2)

y-momentum
$$\frac{\partial q}{\partial t} + \frac{\partial}{\partial y}\left(\frac{q^2}{h}\right) + \frac{\partial}{\partial x}\left(\frac{pq}{h}\right) + gh\frac{\partial \zeta}{\partial y} + \frac{gq\sqrt{p^2+g^2}}{C_2 h_2} -$$
$$\frac{1}{\rho_w}\left[\frac{\partial}{\partial y}(h\tau_{yy}) + \frac{\partial}{\partial x}(h\tau_{xy})\right] - \Omega p - fVV_y + \frac{h}{\rho_w}\frac{\partial}{\partial y}[p_a] = 0$$
(Equation 2.3)

Table 2.1: Symbol list of equations

Symbol		Symbol	
h(x,y,t)	water depth (m)	Ω(x,y)	Coriolis parameter (s^{-1})
ζ(x,y,t)	Surface elevation (m)	p$_a$(x,y,t)	atmosphere pressure (kg/m^3)

$p,q(x,y,t)$	flux densities in x and y directions (m^3/s/m)	ρ_w	density of water (kg/m^3)
$C(x,y)$	Chezy resistance (m$^{1/2}$/s)	x,y	space coordinates (m)
G	gravity acceleration (m/s^2)	T	time (s)
V,V_x,V_y (x,y,t)	wind speed component in x and y direction (m/s)	$\tau_{xx}, \tau_{xy}, \tau_{yy}$	components of effective shear stress
$f(V)$	wind fraction factor		

Roughness coefficients represent the resistance to flood flows in channels and flood plains. The results of Manning's formula, an indirect computation of stream flow, have applications in flood-plain management, in flood insurance studies, and in the design of bridges and highways across flood plains.

Manning's formula is:

$$V = \frac{1}{n} R^{2/3} S_e^{1/2}$$

(Equation 2.4)

where:

V =mean velocity of flow, in meters per second
R =hydraulic radius, in meters
S_e =slope of energy grade line, in meters per meter.
n =Manning's roughness coefficient.

There is a tendency to regard the selection of roughness coefficients as either an arbitrary or an intuitive process. Specific procedures can be used to determine the values for roughness coefficients in channels and flood plains. The Manning's n values for channels are determined by evaluating the effects of certain roughness factors in the channels. Two methods are generally used to determine the roughness coefficients of flood plains. One method, similar to that for channel roughness, involves the evaluation of the effects of certain roughness factors in the flood plain. The other method involves the evaluation of the vegetation density of the flood plain to determine the n value. This second method is particularly suited to handle roughness for densely wooded flood plains. Photographs of flood plains that have known n values confirmed by modelling are presented for comparison with flood plains that have unknown n values.

Setting the Manning's roughness coefficients for flood plains follows a standard approach given in MIKE 21. A Manning number (M) of 30 (n=0.033) has been found to be a practical starting value for most floodplain applications. By altering Cowan's (1956) procedure which was developed for estimating n values for channels, the following equation can be used to estimate n values for a flood plain (Arcement, 1984):

$$n=(nb +n1 +n2 +n3 +n4)m$$ (Equation 2.5)

where nb is a base value of n for the flood plain's natural bare soil surface, n1 is a correction factor for the effect of surface irregularities on the flood plain, n2 is a value for variations in shape and size of the flood-plain cross section, (assumed in this thesis to be equal to 0.0), n3 is a value for obstructions on the flood plain, n4 is a value for vegetation on the flood plain and m is a correction factor for sinuosity of the flood plain, assumed equal to 1.0.

The values for nb, n1, n2, n3, n4 and m can be determined from Table 2.2.

Table 2.2: The adjustment values for factors that affect the roughness of floodplains (Arcement, 1984).

Flood-plain Conditions	n Value adjustment	Degree of irregularity (n1)
Smooth	0.000	Compares to the smoothest, flattest flood-plain attainable in a given bed material.
Minor	0.001-0.005	Flood plain slightly irregular in shape. A few rises and dips or sloughs may be more visible on the flood plain.
Moderate	0.006-0.010	Have more rises and dips. Sloughs and hummocks may occur.
Severe	0.011-0.020	Flood plain very irregular in shape. Many rises and dips or sloughs are visible. Irregular ground surfaces in pasture land and furrows perpendicular to the flow are also included.
Flood-plain Conditions	n Value adjustment	Variation of flood-plain cross section (n2)
Gradual	0.000	Not applicable
Flood-plain Conditions	n Value adjustment	Effect of obstruction (n3)
Negligible	0.000-0.004	Few scattered obstructions, which include debris deposits, stumps, exposed roots, logs, piers, or isolated

		boulders, that occupy less than 5 percent of the cross-sectional area.
Minor	0.040-0.050	Obstructions occupy less than 15 percent of the cross-sectional area.
Appreciable	0.020-0.030	Obstructions occupy from 15 percent to 50 percent of the cross-sectional area.

Flood-plain Conditions	n Value adjustment	Amount of vegetation (n4)
Small	0.001-0.010	Dense growths of flexible turf grass, such as Bermuda, or weeds growing where the average depth of flow is at least two times the height of the vegetation; supple tree seedlings such as willow, cottonwood, arrow-weed, or salt cedar growing where the average depth of flow is at least three times the height of the vegetation.
Medium	0.010-0.025	Turf grass growing where the average depth of flow is from one to two times the height of the vegetation; moderately dense steamy grass, weeds, or tree seedlings growing where the average depth of flow is from two to three times the height of the vegetation; brushy, moderately dense vegetation, similar to 1-to-2-year-old willow trees in the dormant season.
Large	0.025-0.050	Turf grass growing where the average depth of flow is about equal to the height of the vegetation; 8-to-10-years-old willow or cottonwood trees intergrowth with some weeds and brush (none of the vegetation in foliage) where the hydraulic radius exceeds 0.607 m or mature row crops such as small vegetables, or mature field crops where depth flow is at least twice the height of the vegetation.
Very Large	0.050-	Turf grass growing where the average

	0.100	depth of flow is less than half the height of the vegetation; or moderate to dense brush, or heavy stand of timber with few down trees and little undergrowth where depth of flow is below branches, or mature field crops where depth of flow is less than the height of the vegetation.
Extreme	0.100-0.200	Dense bushy willow, mesquite, and salt cedar (all vegetation in full foliage), or heavy stand of timber, few down trees, depth of reaching branches.
Flood-plain Conditions	n Value adjustment	Degree of meander(m)
	1.0	Not Applicable

2.3   Modeling approaches

Several modeling approaches have been adopted in this study. All of them are related to urban flooding and urban drainage systems. This Section discusses the concept and modeling approaches that have been used in this study. Combined sewer systems and storm water sewers in the urban drainage system are designed to relate the rainfall-runoff with a certain designed return period. Under extreme rainfall conditions, flooding can occur at specific locations because of an insufficient capacity of the sewers. Usually, in order to assess the flood risk, a fully dynamic one-dimensional (1D) model is used. Such a model is able to predict accurately flood levels and discharges in applications where the basic assumptions of 1D flow remain valid. This conventional modeling practice however, fails because it is assumed that the water that rises above the ground level and causes flooding which remains at the specific location where it leaves the sewer system.

The local topography is always neglected, except to refine estimates of flood storage above ground. This does not correspond to the reality. It is not only a problem of correctly estimating the local water levels, but in practice water can flow over the streets to lower locations and lead to damage at completely other locations than those predicted with this kind of model. In order to increase the efficiency of the conventional models, the street runoff can be incorporated in the model as a second 1D layer or by using an existing commercially fully two-dimensional (2D) depth averaged flow model. The overall picture of modelling approaches for urban drainage is shown in Figure 2.3.

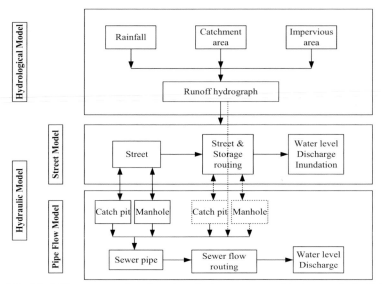

Figure 2.3: Modelling approaches for urban drainage system (Apirumanekul, 2001)

2.3.1 1D modeling

In a 1D model, the network is composed of pipes and manholes. Pipes or open channels are represented by links while manholes are nodes that can also represent basins and outlets. The hydrodynamics of this model are represented by the Saint Venant equation, which take into account conservation of mass and momentum, generating equation for computation to calculate the water levels and the flows on the street and in the sewer pipes. When the water level at node reaches ground level, a virtual reservoir is filled as shown in Figure 2.6. The storage of water surface is not linked to the real capacity of the topography and the flow on the street is not necessarily presented. This situation can lead to an overestimated water depth in the reservoir.

Figure 2.4: Basic 1D model

2.3.2 1D/1D modeling

The 1D/1D coupled hydraulic model solves simultaneously the continuity equation for the network nodes, the complete St. Venant equations for the 1D sewers and the 1D surface networks, and the links equations. The coupled hydraulic model can use the Preissman four-point implicit finite-difference scheme, which can be set to be unconditionally stable.

The multiple-linking-element (MLE) links the 1D sewer network to the 1D surface network. This element is designed for modeling the complex interaction between the sewer and the surface networks. The MLE determines the discharge exchange based on the flow characteristics. The discharge for a given single-linking-element (SLE) link j at a given time k is a function of five control sections (Leandro et al. 2007):

CS1 – From the gutter to the inlet

$$Q_{CS1} = Cd \times \frac{2}{3} H1 \times Li \times \sqrt{\frac{2}{3} g \times H1}$$

(Equation 2.6)

CS2 and CS4 – From the inlet to the vertical pipe and vice-versa (CS4)

$$Q_{CS2} = Cd \times Ap \times \sqrt{2g(H1 + Hi)}$$

(Equation 2.7)

$$Q_{CS4} = -Cd \times Ap \times \sqrt{2g(H2 - H1 - Hi - Hs)}$$

(Equation 2.8)

CS3 and CS5 – From the orifice to the manhole and vice-versa (CS5)

$$Q_{CS3} = K \times Ap \times R^{2/3} \times \sqrt{\frac{(H1 + Hi + Hs - H2)}{Lp}}$$

(Equation 2.9)

$$Q_{CS5} = -K \times Ap \times R^{2/3} \times \sqrt{\frac{(H2 - H1 - Hi - Hs)}{Lp}}$$

(Equation 2.10)

where $Qcsi$ is the flow through the control section i
$H1$ is the water depth at the surface
$H2$ is the water depth at the sewer
Hi is the inlet box height
Hs is the height of the vertical shaft connecting the inlet box to the manhole
Lp is the length of the vertical shaft
Ap is the area of the connecting pipe

Cd is the discharge coefficient

K is the Strickler roughness coefficient

The discharge of the SLE is determined by:

$$Q_k^j = \max\{Q_{CS1}, Q_{CS2}, Q_{CS3}\}$$ (Equation 2.11)

$$Q_k^j = \min\{Q_{CS4}, Q_{CS5}\}$$ (Equation 2.12)

Equation (2.11) is used if the flow is from the surface to the sewer, and Equation (2.12) is used if the flow is from the sewer to the surface. The MLE is obtained as the product of the number of connections (SLEs) to each manhole. In general, the MLE is function of four main variables;

$$\left(Hs_k^j, Hp_k^j, Neq^j, Rcd^j\right)$$ (Equation 2.13)

where Hs_k^j and Hp_k^j are the dependent variables, the water levels at the surface nodes and at the sewer pipe are two parameters of the MLE, respectively the number of equivalent Single-Linking-Elements (SLE) and the coefficient to reduce instability. The links in the 1D surface-network, other than the 1D open channels, are calculated using a weir equation, which is a function of four variables (Djordjevic et al., 2004)

$$\left(Hs_{K.D}^j, H_{k.U}^j, Z_{level}^j, w^j\right)$$ (Equation 2.14)

where $Hs_{K.D}^j$ and $H_{k.U}^j$ are the dependent variables, water level downstream and upstream of the link j, at the surface nodes. Z_{level}^j is the crest elevation considered and w^j is the weir crest width.

2.3.3 1D/2D modeling

1D/2D linked models use 1D unsteady flow calculations to simulate flow in pipes, channels, culverts and other defined geometries, and the 2D calculations are used where the flow is truly two dimensional. The 1D model is the same link-node system used in traditional 1D models, and the 2D domain is defined by a grid of cells with defined slope and roughness. The 1D nodes are linked to a 2D cell. This linking can be static or dynamic. Figure 2.5 shows the 1D and 2D elements in an urban area where overland flow enters an underground drainage network via inlets. During storm events, excess flow may pond in the vicinity of an inlet or may move overland to an adjacent inlet. In the drainage network, obstructions or inadequate capacity may cause the flow to backup or reverse direction.

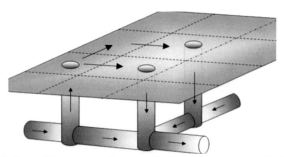

Figure 2.5: Overland flow (2D) and underground drainage network (1D) in a 1D/2D linked model.

Figure 2.6 shows the 1D and 2D elements in an area drained by an open channel. The channel is represented by the cross sections typical of a 1D unsteady flow model while the flood plain is represented as a grid which is the DTM. The elevations of 1D nodes represent the top of the bank of the 1D channel. The surface area of the 1D model is cut out of the 2D grid so that a double accounting of flow volume is not permitted. The water surface in 1D is allowed to rise vertically in the channel and this same water level is tied to the water level in adjacent water cells as defined in the 1D/2D model.

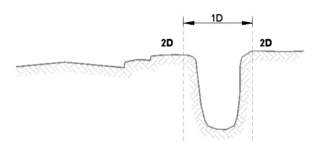

Figure 2.6: Section of channel showing 1D and 2D elements in a 1D/2D linked model

2.4   Modelling uncertainties

Models may be uncertain due to many reasons that commonly relate to the selected model structure, and the implementation of the numerical scheme model input/output and parameters. In flood modelling, with the advance of fast computing facilities and better modelling techniques, it is now possible to solve the full set of equations rapidly and accurately, which can reduce uncertainty.

Selected model structure
A large number of model structures have been developed to predict flood inundation. However,

each structure only approximates nature and therefore has to make many simplifications. For example, unsteady flow component in 1D flood model assumes:

    i.   the flow can be represented by a cross-section mean velocity

    ii.   the water surface is horizontal across any channel section

    iii.   all flows are gradually varied with water of uniform density and hydrostatic pressure prevails at all points in the flow so that vertical acceleration can be neglected

    iv.   the model in this form cannot represent any complex interactions between channel and floodplain and complex floodplain flows

    v.   the channel boundaries are fixed and therefore no erosion or deposition can occur;

    vi.   the resistance to flow under dynamic flow conditions can be approximated by empirical uniform flow formulae such as Manning's or Chezy's equation

These assumptions may not always be fully valid which may also not be necessary in every case. Nevertheless, they will introduce uncertainty in the model results to some extent.

Implementation of the numerical scheme

Most flood models use the St. Venant equations which can be solved with an implicit finite difference scheme using a modified Newton–Raphson iteration technique. The choice of the type of the numerical solution can introduce additional uncertainty in some cases. The user must be aware of the fact that the solution must balance numerical accuracy and computational robustness. Barkau (1997) suggested that larger values of discharge produce a more robust simulation at the cost of accuracy. Vojinovic and Abbott (2011) explained various sources of uncertainty associated with numerical modelling work (e.g., uncertainty in numerical solutions for modelling pressurised flow conditions, uncertainty in the simulation of supercritical and transcritical flows, uncertainties due to model simplification, uncertainties related to system interactions, uncertainties in model calibration – with particular reference to 2D models, etc.).

Another factor which has a great impact on the model robustness is the time step. One rule of the thumb, suggested by Barkau (1997) to ensure proper resolution of the inflow, is to choose a step which is at most one 20th of the time of the rise of the inflow hydrograph. It is also advised that the time step should satisfy the Courant condition, which indicates the limit for explicit numerical solutions, and is satisfied by the followed equation;

$$\Delta x \geq C \Delta t$$

(Equation 2.15)

where Δx = Space step

C = Speed of flood wave

Δt = Time step

However, it remains unclear how to choose the optimal time step, especially because it is reported that for implicit solutions the Courant condition does not need to be satisfied and larger time steps can produce good results (Barkau, 1997). The Courant condition and the analysis of the best weighting parameter are based on the theoretical linear analyses, and in practice many other factors influence the results, including, for example, changes in cross section properties, hydraulic structures and sudden increase/decrease of the channel slope or the Manning roughness. The choice of both time step and weighting parameter can introduce numerical errors, which may be hard to distinguish from responses due to other sources and thus this choice introduces additional uncertainty.

The strengths of flood model structure derive from its explicit dependence on a regular gridded digital elevation model (DEM) to parameterize flows through the riparian areas. Applications for simulating flood wave propagation and inundation over long (101-102 km) river reaches have typically employed grid resolutions between 25m to 250 m (e.g., Horritt and Bates, 2001). At this resolution digital elevation data are widely available from a variety of sources, including digitized contour data and aerial stereo-photogrammetry. However, the use of such coarse spatial discretisations may lead to poor model performance where detailed floodplain geometry, below the resolvable level of the model grid, has a significant effect on flow routing. For finite element models, the use of low resolution discretisations is primarily forced by computational constraints, so that practical applications are limited to the simplified boundary geometries typical of rural and semi-rural floodplains. The influence of sub-grid topography and obstructions such as vegetation and sparse buildings are then treated relatively crudely by incorporating variability in weakly constrained roughness parameters (e.g., Horritt, 2000; Mason et al., 2003).

Until the advent of survey technologies such as LiDAR, computational flood hydraulics was increasingly limited by the data available to parameterize topographic boundary conditions rather than the sophistication of model physics and numerical methods. New distributed data streams such as LiDAR, now pose the opposite problem of how to optimally use their vast information content within a computationally realisable context. An increasingly popular approach to manage this scale problem is the development of subgrid parameterization methods. These have been applied in various forms in hydraulic modelling, to model the friction coefficient of vegetation or as part of a more complex scheme to modify the full shallow water equations to account for small-scale ground irregularities (Defina, 2000). Using this approach, dense urban surface models is used to parameterize a sub-grid model of the flow boundary geometry and then embedded within a coarse resolution model to reflect the first order influences on flow conveyance due to 'blocking' by buildings and micro-topography, reducing floodplain storage and realigning flow paths. This is then used to adjust the continuity equation and update estimates of flow depth and cross-sectional area. This approach can add uncertainty of computational overheads but it can enable simulations at more rapidly realized coarser resolutions.

Parameters

Another factor is parameter uncertainty, which in the flood modeling context usually means surface roughness. It has to be emphasized that uncertainty in effective local surface roughness may influence the inundation pattern at the local scale significantly, but it can only have a small impact on the overall inundation predictions. Furthermore, it will be significantly more difficult to identify parameter combinations in a high dimensional space.

Input/output

Another problem is the impact of input and output uncertainty on the modeling process. Model input is usually a discharge hydrograph, which is either derived from rating curves or is the output of another model. In both cases, significant uncertainty has to be considered. Flood inundation models produce both hydrographs and inundation depths and extent as model output. The latter is, to an increasing degree, compared to the event data which are acquired by manual survey or remote sensing. Significant uncertainties can be introduced into this data due to vegetation, wind or survey errors.

2.5 Data uncertainty

In the context of a flood model, data uncertainty mostly comes from the topography. The topography is one of the main inputs into most flooding models; yet topography is very often seen as the factor with the least uncertainty. Various studies have shown that small errors in the topography can have significant effects on the model results. The issue is made even more complex when the representation of infrastructure is included, though very often it is neglected. In recent years, high resolution DTM has been widely used in automated hydrological analysis. The sources of high resolution DTM are extensive due to the advancement of remote sensing techniques. The derived hydrological feature information is of great help in many hydrological models; some of these features are invaluable for estimating flood extent and timing, surface water runoff calculations, and the prediction of stream discharges. All hydrologic models ultimately rely on some form of overland flow simulation to define drainage courses and watershed structure. To create a completely connected and labeled drainage network with the watershed divide, the water outflow from each cell in the DTM has to be routed to the catchment outlet at the edge of the DTM. The degree of uncertainty in the DTM increases with the increase in resolution of the data.

The uncertainties that influence the derivation of hydrological features include DTM measurement errors, estimation of topographic parameters, the effect of the DTM scale imposed by the grid cell resolution, any DTM interpolation, and terrain surface modification used to generate hydrologically-viable DTM surfaces (Wechsler, 2006). Provided the preprocessing of DTM is perfect, artifacts (spatially structured elevations) and sinks (surface depressions) are the only key elements to be defined. In general, the influence of artifacts and sinks on derived hydro-features is high. They are treated as nuisance features in hydrologic modeling. The common practice is to locate and remove these features in the DTM at the very first step of the hydrologic analysis.

2.6   Current research concerning urban flood modeling practice

Flooding using a 1D drainage model has been represented historically by the use of flood cones at manholes. As the water level exceeds the manhole cover, it continues rising into a conical vessel. This representation of flooding is a simplification which takes no account of the topography around the manhole, which may allow overground flow and reentry into the 1D system elsewhere. From about 2004, many researchers have sought to improve urban flood modelling through the coupling technique. Conversely 2D modeling packages are able to cope well with flood flow over the ground surface but of course are not suitable to model a 1D drainage network. Hence the integration of a 2D model with a 1D model has enabled flooding due to inadequate capacity of a drainage network to be modeled more accurately.

2.6.1   Current research on 1D Flood Model

Apirumanekul C. (2001) studied urban flooding in Dhaka City and tried to find the best solution to mitigate the flooding problem by determining the drainage system capacity and finding a suitable alleviation scheme. The study was based on a physically based 1D model, MOUSE, which was used to simulate the urban drainage system in Dhaka City. A hydrodynamic model was built for two networks, namely, the street and pipe networks. The model describes the flows in both the pipes and the streets together with the flow exchange between these two systems. The results from the model are presented by flood inundation maps in a GIS.

Iwata et. al., (2001), simulated flood in highly urbanized areas of Japan to make hazard maps based on flood simulations, as a counter measure to prevent serious damages. The study characterized the flooding mechanism in urbanized areas in three constituents: the flow on ground surface, the flow in the drainage system, and the flow in underground drainage; all were modeled using MOUSE. Also two types of flooding were recognised. One is caused by precipitation and the other is generated by overflows from a river, which can produce more severe damages. The study suggested that the river network should be added to the models in order to obtain more accurate results. Therefore, a comprehensive simulation model of flooding in a highly urbanized area should consist of five sub-models (a rainfall-runoff model, a drainage system model, an underground drainage modela road network model and a river network model). Such a model is prepared in this thesis to study the danger of inundation in urban areas and the effects of countermeasures.

Rodolfo A, (2003) reported that the city of Buenos Aires frequently suffers from serious flooding that damages property, the economy and brings misery to the lives of the city's 3.5m inhabitants. Over 25 deaths have been recorded since 1985. A two-year study funded by the World Bank is being focused on the Buenos Aires drainage system. This is helping to identify structural weaknesses in the surface and underground networks. The choice of InfoWorks CS as the modelling package for this major study was very specific. The main reason is because of its database structure, GIS integration, simulation performance and its SQL Server and Oracle links. Other aspects like real time control (RTC) were also important. The model contains approximately 18,000 nodes, 12,000 on street level and 6,000 underground. 25,000 pipes are detailed of which 18,000 are used to model overland flow at street level, while the remainder represent the actual underground sewer system. The model contains 7,000 sub-catchments. One of the main challenges of the study was to come up with an accurate diagnosis of water behaviour at street level in order to predict accurately the effect of runoff on the underground drainage system. Each street has unique conveyance characteristics, and for flood level predictions to be accurate, the flow paths of the flood water must be traced precisely. The study has revealed both the lack of capacity in the underground system and the lack of inlet capacity in some localized areas. Structural adjustments in both aspects can help alleviate the effects of flooding.

Suresh L. et. al., (2004) reported that the city of Minneapolis hired HDR to provide consulting engineering services to help analyse and develop solutions to the problems residents face in flood zones. A model of the storm sewer network was developed using the XP-SWMM modelling software in order to analyse the existing system and develop possible solutions. The storm sewer model was refined to evaluate timing impacts on a creek. The watershed discretization was for 30 sub watershed. The streets above the main sewer lines that convey excess runoff were added to the model to simulate accurately the flooding at low area. The proposed solution from this study will protect homes during an extreme storm (defined as the 100-year storm for this location). The approach allows water to pond on the street only to a level that will not cause damage to homes.

Mark et. al., (2004), outlined modeling approaches and principles for analysing urban flooding. They describe how urban flooding can be simulated by one-dimensional hydrodynamic modelling incorporating the interaction between the buried pipe system, the streets (with open channel flow) and the areas flooded with stagnant water. According to the authors, the modeling approach is generic in the sense that it handles both urban flooding with and without floodwater entry into houses. In order to visualize the flood extent and impact, the modeling results are presented in the form of flood inundation maps produced with the GIS. In this paper, only flooding from local rainfall was considered together with the impact in terms of flood extent, flood depth and flood duration. The authors also discuss the data requirements for verification of urban flood models, together with an outline of a simple cost function for estimating the cost of the flood damages.

Recently, the Klang river basin has introduced a decision support system for flood forecasting.

This system is called FLOOD WATCH and is based on a MIKE 11 hydrodynamic model and the MIKE 11 FF real-time forecasting system. This system is integrated with GIS. The FLOOD WATCH system has been developed and implemented within the ArcView GIS environment. With recent advances in GIS and computer technology; flood forecasting has improved considerably. FLOOD WATCH is a powerful tool for real-time flood forecasting and flood warning. But before the system can be used, databases have to be established and graphical displays of the stations have to be configured. The system can run in automatic mode or in manual mode in which the operator controls the system. There are four main processes in FLOOD WATCH:

i.   Input –telemetry data is imported from the hydro-meteorological network.

ii.   Pre-processing – this is done using tools for interpolation of input data, making a consistency check of data, filling in missing input simulation data and performing Q/H calculation and data shifts

iii.   Modelling – once a request for a forecast is made, the system automatically extracts the required data from the FLOOD WATCH database to MIKE 11. Model simulations are executed in MIKE 11 and the results of the simulated forecasts are transferred to the FLOOD WATCH database for display and further dissemination.

iv.   Output – the graphical display in Arc View is automatically updated with the most recent flood information. Graphs of measured and forecast water levels and discharges can also be derived from FLOOD WATCH.

Boonya A.S et. al., (2007) has presented an innovative method for the analysis of the overland flow component during pluvial flooding in urban areas. The concept is based on access to a detailed high quality DTM, from which the surface network of ponds and preferential paths between manholes is created. The surface network interacts with the underground sewer network. This concept appears to be a significant in terms of a global scale approach to urban flood modeling. It is hoped that this development will motivate professional software companies to follow this approach. Future research will include objective comparisons between the present methodology and coupled 1D/2D models. The work presented in this research is the first (though very important) phase in enabling a fully integrated urban pluvial flood to be modeled using the 1D/1D approach. It should be noted that success in implementing this concept depends on progress in improving the vertical resolution of the DTM for complex urban areas. Results from the study open up several new areas of advanced urban flood management including improvements in Real Time Control, links with rainfall now-casting, and the development of short term urban flood forecasting.

2.6.2 Current research on 1D/2D Flood Model

Md. Jahangir Alam (2003) studied the capability of a 2D urban flood modeling approach and compared it with a 1D flow modeling technique. The study was based on coupling MOUSE for the pipe flow model and MIKE 21 for the surface flow model. In additional, a real time urban flood model for Dhaka city was also developed for flood forecasting and real time control. The model results are presented in GIS by the development of flood inundation maps. It is stated in the report on the research that this coupling approach provides good information for flood risk and damage assessment.

Mark et. al., (2004), outlined the modeling approaches and principles for the analysis of urban flooding. They describe how urban flooding can be simulated by one-dimensional hydrodynamic models incorporating the interaction between the buried pipe system, the streets (with open channel flow) and the areas flooded with stagnant water. According to the authors, the modeling approach is generic in the sense that it handles both urban flooding with and without flood water entry into houses. In order to visualize flood extent and impact, the modeling results were presented in the form of flood inundation maps produced in GIS. In this paper, only flooding from local rainfall was considered together with the impact in terms of flood extent, flood depth and flood duration. The authors also discussed the data requirement for verification of urban flood models together with an outline of a simple cost function for estimation of the cost of the flood damages.

T.G. Schmitt et al (2004) also looked at the modeling of flooding in urban drainage systems. In their study, the modelling system RisUrSim was developed in order to meet the requirements of simulating urban flooding, focusing on the occurrence of separate flow over the ground surface and its possible interaction with the surcharged sewer system. RisUrSim consists of three sub-modules: rainfall-runoff (Risoreff), surface flow (Risosurf) and sewer flow (HamokaRis). The module Risoreff is applied to areas where surface flooding and the interaction between surface flow and sewer flow can be excluded. The module Risosurf computes surface flow based on a simplified representation of the shallow-water equations using GIS-based surface data. HamokaRis is applied for all underground drainage elements. RisUrSim allows bi-directional exchange of flow volume between surface flow module Risosurf and the sewer flow module HamokaRis at defined exchange nodes. The implementation of the coupling between Risosurf and HamokaRis requires a careful consideration of numerical stability and the conservation of water. Numerical stability is secured with a synchronized administration of the seclection of the appropriate time steps. RisUrSim has been applied in the city of Kaiserslautern, Germany to prove the concept of the method. The authors state that with the RisUrSim model, the surface flooding could be reproduced realistically.

Vojinovic et. al., (2006), discussed the potentials and limitations of one- and two dimensional physically-based modeling approaches for describing flow transitions at street junctions. From the analysis of model results, they conclude that for modeling sub-critical flow conditions the difference in 1D and 2D model results is not so significant when compared to the discharges and water levels. However, for modeling supercritical conditions, the difference between 1D and 2D models is more significant in that the 1D model misrepresents both the discharge and the water level values. This is due to the simplified calculation performed at the nodes of the 1D model. They highlight the need to apply 2D models for modeling street junctions in urban areas.

NguyenV.D et al (2008) studied how to improve flood inundation modeling in inundated areas in the Mekong Delta using integrated data from remote-sensing, in-situ measurements and from multi-objective global optimization techniques. The authors develop first a 1D model for the whole Mekong Delta (MODEL A). Then they produced a 1D model (MODEL B-1D) for the river network in the study area which is more detailed than MODEL A. They used MIKE 11 and a 2D model (MODEL B-2D) for the floodplain in the study area using MIKE 21. These 1D and 2D models are coupled to form MODEL B using MIKE FLOOD. Results from MODEL A for the whole MK Delta are used as inputs for the coupled model. The final results show that the use of in-situ data from MODEL A, a parallelization process for grid computing and multi-objective optimization, increases the performance of the coupled model.

George (2008) outlines some initial testing undertaken on the new InfoWorks CS model which uses a 2D overland flow model instead of the traditional 1D flood cones. He outlines a novel approach for generating rural runoff on the 2D mesh. In addition, some of the issues associated with linking the 1D and 2D models and the modeling of open channels are discussed. HR Wallingford has conducted research into the application of InfoWorks 2D to modeling flooding on a site with a particular set of challenges. The goal was to produce a model which can be used to simulate extreme (10,000 years) events through to moderate flood events successfully. In building a model using InfoWorks 2D, a number of different approaches have been taken in modeling rural runoff, river channels and piped drainage networks. The approaches are divided into 3 main themes:

i. Generation of rural runoff using a grid of sub catchments with associated 2D nodes covering the rural area so that the correct flows with their appropriate timings enter the site. Upstream catchment runoff has been placed directly onto the 2D mesh. This is then routed by the mesh topography towards the site itself

ii. River channels convey a proportion of the total flood flows in the rural catchment; hence they are included in the model. The proportion of river flow decreases with the increasing of flood severity.

iii. The site drainage networks are included in the model to correctly simulate the effect of the drainage system during flood events. Again, any piped system has a decreasing effect with longer return period storms.

The initial testing, which went into the development of this approach, has proved extremely useful in understanding the capabilities of this cutting edge software. The approach represents a very considerable improvement on the previous 1D simulations and gives a much greater degree of flexibility than other 2D modeling packages.

Bryan E. et al (2009) come up with a 'modeling approach to support the management of flood and pollution risks for extreme events in urban stormwater drainage systems'. In this study, the authors state that the complexity and the interactions of flooding and associated pollution resulting from exceedance of surface water flows during extreme storm events present a major threat for future sustainable urban drainage management. Innovative coupled 1D/2D modeling approaches are used for the detailed delineation and analysis of surface flow paths, flood depths and velocities during such extreme events. The modeling structure and outputs are referred to as the Birmingham Eastside SWITCH demonstration area. From the results, the 1D/2D approach provides a realistic analysis of extreme event overland surface flows, especially where the flow paths are not confined to preferred street sections. The outputs allow the testing of different flood planning scenarios and the generation of risk/hazard maps as well as enabling a flood vulnerability analysis. However, the modeling methodology is very sensitive to the DTM data and allocation of roughness values, requires small scale time steps and is also computationally demanding. As such it is probably inappropriate for real time flood representation and rapid forecasting, being much better suited as a planning and management screening tool for identification and quantification of flood areas, depths and flow paths. This makes it suitable for selecting locations for mitigating measures and deciding on emergency/hazard planning procedures, as well as a basis for analysing a detailed depth-damage costing.

Chapter 3

TWO DIMENSIONAL (2D) SURFACE MODEL

3.1 Introduction

The surface model is a digital representation of the ground surface topography or terrain. A surface model can be represented as a raster map, in the form of a grid of squares or as a triangular irregular network. Surface models are often used within a GIS, and are the most common basis for generating digital relief maps.

3.2 Concept of a 2D surface model

Digital Terrain Models (DTM) may be prepared in a number of ways, but they are frequently obtained by remote sensing, rather than from a direct survey. One powerful technique for generating digital elevation models is by using LiDAR.

Older methods of generating DTMs often involve interpolating digital contour maps that may have been produced by a direct survey of the land surface. A DTM implies that the elevation is available continuously, at each location, in the study area.

The quality of a DTM is a measure of how accurate the elevation is at each pixel (absolute accuracy) and how accurate the morphology is presented (relative accuracy). Several factors play an important role in the quality of DTM-derived products:

   i.   Terrain roughness
   ii. Sampling density (elevation data collection method)
   iii. Grid resolution or pixel size
   iv. Interpolation algorithm
   v.   Vertical resolution
   vi. Terrain analysis algorithm

3.3 Types of 2D surface models

There are two types of 2D surface models that can be used to represent the ground surface, namely Digital Terrain Model (DTM) and Digital Surface Model (DSM).

3.3.1 Digital Terrain Model (DTM)

Perhaps out of all necessary data, a DTM provides the most essential information nowadays for flood managers. A DTM refers to a topographical map, which contains terrain elevations, and as such, it is used to represent the terrain (or land) surface and its properties. A DTM generally refers to a representation of the Earth's surface (or a subset of this), excluding features such as vegetation, buildings, bridges, etc. (refer to Figure 3.1). The DTM often consists of raw data. Correspondingly, the term DEM (Digital Elevation Model), although usually associated with the land surface, refers to elevations of any type of surface.

In urban flood management, DTMs are needed for the analysis of terrain topography, for setting up 2D models, processing model results, delineating flood hazards, producing flood maps, estimating damages, and evaluating various mitigation measures. Typically, a DTM data-set can be obtained from ground surveys (e.g., total stations, together with Global Positioning System (GPS), aerial stereo photography, satellite stereo imagery, airborne laser scanning, or by digitising points/contours from an analogue format (e.g., a paper map to digital format), so that they can be stored and displayed within a GIS package, and then interpolated.

A DTM can be represented as a raster map (a grid of squares) or as an irregular triangular network. DTMs are often used in geographical information systems, and are the most common basis for digitally-produced relief maps. A DTM is often required for flood or drainage modelling, land-use studies, geological applications, and much more.

Figure 3.1: Example representation of a DTM

3.3.2 Digital Surface Model (DSM)

A Digital Surface Model (DSM) on the other hand, includes buildings, vegetation, and roads, as well as natural terrain features (Refer to Figure 3.2). A DSM may be useful for landscape modelling, city modelling, and visualization applications.

Figure 3.2: Example representation of a DSM

3.3.3 Sub-grid model

Producing a realistic surface for water flow patterns can be difficult for hydrologic models when there is insufficient grid resolution as a result of computational constraints or when available DTM data is relatively coarse. It is likely that a subgrid-scale treatment of topographic detail will continue to be required.

Yu and lane (2006) explored sub-grid model treatments with a mesh resolution that followed a dyadic series (i.e. $R22n$), for $n = 0$–3, where r is the finest resolution of data available. For each mesh resolution defined by n (except for $n = 0$), they assessed the effects of sub-grid-scale treatments with a resolution of $2n + m$ for $m = 1$. They found that the use of a sub-grid-scale treatment significantly improved model performance, but they did not consider the situation where $m > 1$. Yu and lane (2011) in their paper assess the effects of situations where $m > 1$, as an attempt to begin to identify just how coarse a model resolution can become, whilst still retaining a reasonable representation of the effects of floodplain elements upon the flood inundation process. They include the effects of post-processing of topographic input data upon the representation of floodplain features, as the averaging process used to determine the elevations of cells in coarser meshes can often smooth these features out of the analysis.

3.4 2D surface model approaches

Generally, there are two approaches that are commonly used to represent a 2D surface model. These approaches are a regularly spaced grid and a Triangulated Irregular Network (TIN).

3.4.1 Regularly spaced grid

The most common DTM data structure is raster based (or grid structure). This normally consists of a matrix of square grid cells, with the mean cell elevations stored in a two dimensional array, as shown in Figure 3.3. The location of a cell in a geographical space is implicit from the row and column location of the cell within the array; provided that the boundary coordinates (geo-references) of the array are known. Grid DTMs are widely available and are used because of their simplicity, processing ease, and computational efficiency (Martz and Garbrecht, 1992). Disadvantages include the grid size dependency of certain computed topographical parameters (Fairfield and Leymarie, 1991) and the inability to locally adjust the grid size to the dimensions of the topographical land surface features.

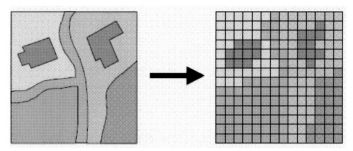

Figure 3.3: Objects represented as a regularly spaced grid

3.4.2 Triangulated Irregular Network (TIN)

A TIN is a digital data structure, used in a Geographic Information System (GIS), for the representation of a surface. A TIN is a vector based representation of the physical land surface or sea bottom, made up of irregularly distributed nodes and lines, with three dimensional coordinates (x, y, and z) that are arranged in a network of non-overlapping triangles (refer to Figure 3.4). An advantage of using a TIN over a raster DTM in mapping and analysis, is that the points of a TIN are distributed variably, based on an algorithm that determines which points are most necessary for an accurate representation of the terrain. Data input is therefore flexible and fewer points need to be stored than in a raster DTM with regularly distributed points. A TIN may be less suited to certain kinds of GIS applications than a raster DTM, such as for the analysis of a surface's slope and aspect.

A TIN is comprised of a triangular network of vertices, known as mass points, with associated coordinates in three dimensions, connected by edges to form a triangular tessellation. Three-dimensional visualizations are readily created, by rendering the triangular facets. In regions where there is little variation in surface height, points may be widely spaced, whereas in areas of more intense variation in height, the point density is increased.

A TIN is typically based on a Delaunay triangulation, but its utility can be limited by the selection of input data points i.e., well-chosen points need to be located such that they capture significant changes in surface form, such as topographical summits, breaks of slope, ridges, valley floors, pits, and cols. In mathematics and computational geometry, a Delaunay triangulation for a set P of points in a plane is a triangulation DT(P) such that no point in P is inside the circumcircle of any triangle in DT(P). Delaunay triangulations maximize the minimum angle of all the angles of the triangles in the triangulation; they tend to avoid skinny triangles (Mark et.al, 2008).

Although usually associated with 3-dimensional data (x, y, and z) and topography, TINs are also useful for the description and analysis of general horizontal (x and y) distributions and their relationships. The first triangulated irregular network program for GIS, was written by Randolph Franklin, at the Simon Fraser University, in 1973.

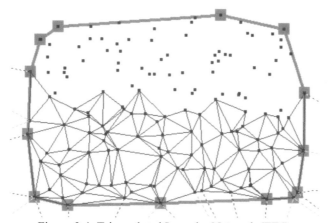

Figure 3.4: Triangulated Irregular Network (TIN)

3.5   Interpolation

Interpolation refers to the process of estimating an intermediate value from two known values. Clearly, this is only an approximation, and as such, care must be taken when using interpolation.

3.5.1 Interpolation algorithms

Linear Interpolation

Generally, if you know the values of $y(x_i)$ and $y(x_{i+1})$, the value of $y(x > x_i, x < x_{i+1})$ can be approximated by:

$$y(x) = y(x_i) + \frac{y(x_{i+1}) - y(x_i)}{(x_{i+1}) - x_i}(x - x_i)$$

(Equation 3.1)

Higher Order Methods

The drawback of linear interpolation is that it is assumed that the function between two known points is a straight line. Some methods have been suggested that try to fit a polynomial, or other known curve to the data, in order to get a slightly better approximation.

Due to the uncertainty that is inherent in any data, this is considered extremely risky unless you are absolutely sure that the functional form you are assuming is correct, and that your data has a very high signal to noise ratio.

3.5.2 Common Interpolation algorithms for DTM interpolation

Some of the most common interpolation algorithms in the industry are based on Inverse Distance Weighted (IDW), Kriging, Spline, and Natural neighbour.

IDW

IDW determines cell values using a linear-weighted combination set of sample points. The weight assigned, is a function of the distance of an input point, from the output cell location. The greater the distance, the less influence the cell has on the output value. For this reason, the IDW function should be used when the set of points is dense enough to capture the extent of local surface variation needed for analysis. A sample result for this interpolation is shown in Figure 3.5.

Figure 3.5: Sample results for IDW

Spline

A spline estimates a value for a variable using a mathematical function that minimizes the overall surface curvature. This results in a smooth surface that passes exactly through the input points. There are two variations of spline, which are regularized and tension. A regularized spline incorporates the slope (first derivative), rate of change in slope (second derivative), and rate of change in the second derivative (third derivative), into its minimization calculations. A sample result for this interpolation is shown in Figure 3.6. A tension spline uses only the first and second derivatives, but includes more points in the calculations.

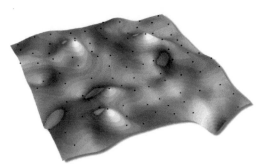

Figure 3.6: Sample results for regularized Spline

Kriging

Kriging assumes that the distance or direction between points reflects a spatial correlation that can be used to explain the variation in the surface. It fits a function to a specified number of points or all points within a specified radius, in order to determine the output value for each location. Kriging is most appropriate when a spatially correlated distance or directional bias in the data is known. The predicted values are derived from the measure of relationship in samples using a weighted average technique. It uses a search radius that can be fixed or variable. A sample result for this interpolation is shown in Figure 3.7.

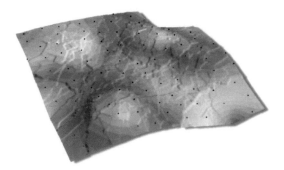

Figure 3.7: Sample results for Kriging

Natural neighbour

Natural neighbour can be used for interpolation and extrapolation. It is another weighted method that generally works well with clustered scatter points. The basic equation used is identical to IDW. This method can efficiently handle large data sets of input points. A sample result for this interpolation is shown in Figure 3.8.

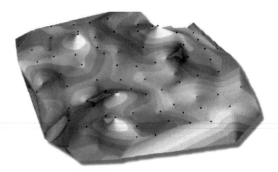

Figure 3.8: Sample results for Natural neighbour

3.6 Representation of an Urban Surface

3.6.1 Urban Surface

Rapid urbanization and accelerated urban sprawl have significantly impacted on the urban climate (Voogt and Oke, 2003) and on urban biophysical processes and the physical environment; they consequently influence the quality of human lives. Timely and accurate information on the status and trends of urban ecosystems and biophysical parameters is critical when developing strategies for sustainable development and for improving the urban residential environment and living quality (Yang et al., 2003; Song, 2005). Therefore, developing techniques and enhancing the ability to monitor urban land use and land cover changes are important for city modelling and planning.

3.6.2 Essential Objects of an Urban Surface to be represented in an urban flood model DTM

Urban environments can contain a vast variety of features (or objects), which certainly have a role in storing and diverting flows during flood events. In this respect, buildings represent one of the most essential objects, and in broad terms, they can be divided into three types: buildings with basements, passage buildings, and buildings which have neither basements nor passages, Figure 3.9. Typical examples of buildings with basements are buildings with underground car parks, which can be subject to flooding, during flood events. Passage buildings refer to buildings that have no basements, but do have large open spaces and corridors that can allow the flow of floodwater throughout the building. The third category refers to those buildings that may act as solid objects, and whose structure will almost fully divert floodwater.

In addition to buildings, there are also many smaller geometric 'discontinuities', such as roads, stairs, pavement curbs, fences, and other objects, which could play an important role in diverting shallow flows that are generated along urban floodplains. In cases where those features are not adequately represented, it is highly likely that the model will not be able to produce satisfactory results.

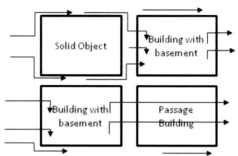

Figure 3.9: Schematisation of water flow through and around different types of buildings.

Elevated roads and bridges are common objects in urban areas and their purpose is to span the gap between two land masses. Such objects, and particularly elevated roads, can cause unrealistic obstruction to the flow of water. In some cities around the world, and particularly in Asia (e.g., Kuala Lumpur, Bangkok, Tokyo etc.) elevated roads are very common and they can occupy significant portion of urban space. Current literature concerning LiDAR data processing shows very little progress in detecting such structures and most of the work to date is based on the use of satellite imagery and radar. Evans (2008) utilized a "fuzzy" based approach for the identification of the bridge edges with a limited amount of available LiDAR data. This methodology allowed for the detection of areas that are deemed likely to be bridge edges. There are also several studies where Synthetic Aperture Radar (SAR) has been used to detect bridges in urban areas (see for example, Houzelle and Giraudon, 1992 and Wang and Zheng, 1998). The principle used in detection of bridges is based on the assumption that every bridge contains the same material as nearby roads. Trias-Sanz and Lomenie (2003) proposed an artificial neural network based approach which can detect bridges in high resolution satellite imagery data using the following assumptions:

i.   Large regions of water or railway yard are separated by a narrow and long strip. This strip is a bridge.

ii.   A small gap between two regions that have been identified as road and are aligned is a bridge. The definition of small is defined by the user.

iii. A small gap between two regions that have been identified as canal and are aligned is a bridge. The definition of small is user defined.

Other authors have used a segmentation method to process LiDAR data. Sithole and Vosselman (2006) have used a segmentation method to detect bridges under the following assumptions:

i.   A bridge is connected to the bare earth on at least two sides.

ii.   To span land masses a bridge will along its length necessarily is raised above the bare earth. Hence, along the length of a bridge, diametrically opposite points on its perimeter are raised above the bare earth.

iii.   A bridge is typically built to be longer than its width.

In order to properly represent elevated roads and bridges in a DTM that is required for 1D/2D flood modelling purposes, two different approaches can be applied. For a normal bridge (i.e., a bridge that crosses a river), it should be completely removed from the DTM as it is normally accounted for in a 1D model. For an elevated road, while this object needs to be removed from the DTM, if the structures underneath (such as piers) occupy a larger area which may cause diversion of the flow, they may still be needed. Currently available filtering algorithms can only partially remove an elevated road from the DTM. In such case, the algorithm would leave behind a non-uniform topography that generates an obstruction that does not necessarily correctly represent the reality under the elevated roads. Furthermore, besides the major structure, there may be structures underneath (such as piles) that need to be considered. As these (lower) structures can play an important role in description of a flood pattern it means that the careful attention should be paid to their treatment and representation.

Street detection on the digital surface model is an important and necessary task for the preparation of a detailed urban surface in order to simulate flooding phenomena in urban areas. A street surface is usually a place where runoff occurs and where sewer and surface water interact with each other. Manholes and water inlets are the gateway that facilitates this interaction. Another important emphasis of this study is the detailing of streets, sidewalks and the height of the curbs.

Street extraction from LiDAR data is not a new idea. In fact some work has been done on this issue. Mainly researchers have adopted two approaches, namely automatic and semi-automatic, to detect street network in the LiDAR data. In automatic approaches, researchers mostly follow the conventional way by converting the LiDAR point cloud into a gray scale image and then applying digital image processing technique on the gray scale image. In semiautomatic approaches the combination of existing vector road maps and a raster LiDAR image is used to detect the road network.

Hinz and Baumgartner (2003) have performed street extraction in complex urban scenes from multi-view aerial images with a high ground resolution. They use a road model exploiting knowledge about the radiometric, geometric, and topological characteristics of roads, making use not only of the image data, but also of a Digital Surface Model (DSM). First, separate lanes are extracted as 2D segments. These lanes are then merged in a fusion process that makes use of the DSM. The street network is constructed in an iterative way from these street segments. The main problems identified were the influence of large vehicles on the extraction process and the weakness of the model at intersections, which affected the linking of lane segments.

Huber and Lang (2001) performed street extraction from high-resolution (2 m pixel size) airborne X-Band SAR data. A street was modelled by three homogenous regions: one central region surrounded by two adjacent regions on either side of the road by way of operator fusion. The result is complemented by using Active Contour Models to very preliminary and the authors concluded that several improvements to the algorithm need to be investigated.

There have been relatively few attempts to extract roads from LiDAR data. Most methods require a form of data fusion to complete the task. Clode et. al (2004) presented a method for the automatic detection of streets from airborne laser scanner data. This study deals with the use of as much of the recorded laser information as possible; thus both height and intensity are used. To extract roads from a LiDAR point cloud, a hierarchical classification technique is used to classify the LiDAR points progressively as road or non-road. Initially, an accurate digital terrain model (DTM) model is created by using successive morphological openings with different structural element sizes. Hatger and Brenner (2003) performed an estimation of road geometry parameters using high resolution LiDAR data (4 points per m2). LiDAR data is used in conjunction with existing database information to derive properties such as height, slope, curvature and width, with a view to using this information in future driver information and warning systems. Although the detected properties are geometric properties of a road, they can only be detected once the lower level geometric properties such as the centreline or road segment have been determined. In this method this information was not extracted: it was provided from an existing database.

In this research a semiautomatic approach for the detection of street is described. A combination of the vector data and raw LiDAR point cloud is used in this approach. This is because raw points preserve the originality of the surface before they are converted to any other formats. Curbs are recovered mostly by detecting streets segments and using the data fusion concept.

Besides man-made objects, natural objects such as vegetation (trees, scrubs, grass etc.) are also taken into account as essential objects to be represented in an urban flood model DTM. Schubert et al. (2008) in their study developed a method to take resistance parameters from the literature and distribute these across a study site, using aerial imagery and LiDAR terrain data as a guide. Their method is inspired by the widespread coverage of aerial imagery and the opportunity to further exploit LiDAR data obtained for the terrain description. In their study, aerial imagery is used to classify typical landcover such as buildings and roads including bare soil, grass and trees/bushes. LiDAR data are processed to estimate a local feature height (e.g., height of shrubs or grasses) which is used to scale the resistance parameter between a minimum and maximum value for each landcover set by the modeler. Consequently, the local resistance parameter becomes a function of the local landcover and the local feature height. It then becomes the task of the modeler to identify a reasonable set of landcover classes and to set minimum and maximum values for each class based on literature values or professional judgment. In this regard, the proposed method may differ significantly from highly mechanistic schemes that limit the need for professional judgment. Using model simulations of urban flooding, the authors evaluate whether resistance parameters based on feature height, in addition to landcover, offer any advantages over resistance parameters based on landcover only. In addition, they evaluate the merits of aerial imagery versus cartographic databases in the context of resistance parameter estimation.

Tyrna and Hochschild (2010) in their study tried to find out if high resolution, QuickBird satellite imagery can be used to model the surface sealing in an urban environment and if this information can be used as an input to calculate surface runoff for storm events with a physically based rainfall-runoff model. Surface sealing means any activity or process in which ground surface areas are packed or plugged to prevent percolation or the passage of fluids. The resulting raster file of the surface sealing model contains information about the pervious and impervious portion of each grid cell. This sub-grid variability is used to calculate spatially distributed excess rainfall values. The widely used SCS Curve Number Method (USDA 1986) was adapted in this study to calculate a pixel-based composite curve number (CCN) for each grid cell according to the following formula:

$$CCN = CN_i * a + CN_p * (1-a)$$

(Equation 3.2)

where a represents the degree of surface sealing reaching from $a = 0$ (no surface sealing) to $a = 1$ (100% impervious). The curve number (CN) is a runoff coefficient which is dependent on land use and hydrologic soil group. For the impervious part of each grid cell, a is multiplied with $CN_i = 98$ as commonly used for impervious areas. According to the LULC classification the prevailing land cover of the pervious parts is grass. The combination of land cover type grass and soil group leads to a specific CN_p value. In this study, the excess rainfall is then calculated with the Curve Number formula using the CCN for each grid cell:

$$Q = \frac{\left(P - \dfrac{5080}{CCN} + 50.8\right)^2}{P - \dfrac{20320}{CCN} - 203.2}$$

(Equation 3.3)

Where P = Rainfall [mm] and Q = Excess rainfall [mm]

3.7   The Influence of DTM Resolution on Urban Flood Modelling Results

In terms of the DTM resolution, it is obvious that a more detailed 2D model (i.e., a 2D model with a higher resolution) will enable better capturing of urban features, whereas such features would be smeared or completely removed, when the grid is coarsened; see Figure 3.10. An illustration of the impact of increases in the 2D model's grid size on buildings and roads is shown in Figure 3.10.

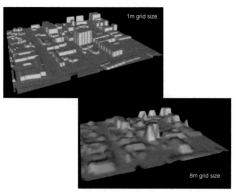

Figure 3.10: Effects of grid resolution on capturing building and road features (Source: Vojinovic et al., 2010 and Vojinovic and Abbott, 2012).

Furthermore, previous research shows that 2D models derived from a coarser DTM resolution, tend to cause a wider spread of floodwater with shallower depths, compared to those models which are derived from finer DTM resolutions; where the flood water tends to get trapped by local terrain depressions, and thus, generates larger depths (Vojinovic et al., 2010); see Figure 3.11.

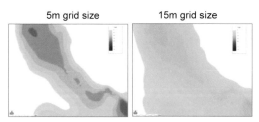

Figure 3.11: Computed water depths for 5m (left) and 15m (right) DTM resolution (See also, Vojinovic et al., 2010b).

It is important to note that current computing resources still impose a challenge for the modelling of large scale areas, where increasing the grid size can effectively improve efficiency, but it can also cause reduction in detail. Again, the straightforward 2D modelling technique may not reflect properly the local flow phenomena in coarse grid applications. In such cases, an approach based on the adjusted conveyance and storage characteristics, may prove beneficial (Seyoum et al., 2010, Vojinovic et al., 2010a). Furthermore, parallel computing is worth exploring; for example, see Neal et al., (2009) and Neal et al., (2010). Lamb et al., (2009) also demonstrate that the use of technology from the computer graphics industry to accelerate a 2D diffusion wave (non-inertial) floodplain model can be beneficial.

Chapter 4

AIRBORNE LASER SCANNING (ALS) AND FILTERING ALGORITHMS

4.1   Introduction

During the last decade, there have been decisive technical improvements made in airborne laser scanning and it has become a standard and well-accepted method for the acquisition of topographic data and for many associated applications. The method is characterised by a far-reaching degree of automation, with digital data recording, and computer-based data analysis. Laser scanning also allows data sampling under meteorological conditions that are too poor for aerial photography, as it uses active illumination (Th. Geist and J. Stötter, 2002). The term 'laser scanning', refers explicitly to a laser system with a scanning capability. This term often refers to a Light Detection and Ranging (LiDAR) installation, carried by an airborne vehicle. LiDAR is a surveying method that can offer us an accurate x, y, and z (height) position. There are several techniques used to obtain height remotely, such as aerial photos and InSAR, but LiDAR offers accurate positioning over large areas which is cost and time effective.

4.2   Airborne laser scanning

Airborne Laser Scanning (ALS) or Light Detection and Ranging (LiDAR) is one of the most common techniques used to measure the elevation of an area accurately and economically in the context of flooding projects involving cost/benefit analysis. It can deliver information on terrain levels to a desired resolution. The end result of an ALS survey is a large number of spot elevations which need careful processing. In this research, the LiDAR System used was Riegl LMS Q560 with Full Waveform Analysis for unlimited target echoes. The Pulse Rate was 75kHz or 75,000 points per second. This system used all echoes and all intensities. The beam size was less than 1.5m diameter with 60 degrees swath width. The data has 40% side lap where the laser distance is approximately 700m. The flying height was approximately 700m above ground level with the average of 100knots flying Speed. The platform for this system was a helicopter (Bell 206b JetRanger). The LiDAR System uses the Differential Global Positioning System (DGPS) with NovAtel's DL-4 data logger which is a high performance GPS receiver with fast data update rates. The GPS antenna receiver is mounted in the aircraft's tail, away from aircraft's rotor and other obstructions to prevent GPS signal interruption and provide highly accurate data.

For the LiDAR data captured in this research, the DGPS base station is located at Subang, Malaysia.

4.2.1 Principle of ALS

ALS is an optical remote sensing technology, which measures the properties of scattered light radiating from of a distant target. LiDAR systems emit highly directional pulses of laser light and measure precisely the time taken for a reflection to return from the ground below. As the speed of light is known (approximately 0.3metres per nanosecond) the distance from the target can be calculated. In addition, the scanner viewing angle at the time each pulse reflection is received, and the sensor's position and orientation are available; therefore, the 3D position of the target can be defined. The sensor position is provided by a GPS/IMU system (Global Positioning System and Inertial Measurement Unit) and the viewing angle is recorded by the sensor itself. Thus, in principle, the positions of the laser hits are determined in-situ, although in practice the final positions are produced during post-processing, when certain sources of bias are removed (Wehr and Lohr, 1999).

Although nowadays, practically all sensors are capable of recording multiple reflections per pulse, a laser pulse cannot penetrate solid surfaces, due to the wavelength used. Therefore, as not every pulse generates multiple returns, only those pulses that encounter features, which at least partially allow light to penetrate them (such as the forest canopy), generate multiple reflections. In simplified terms, the first return is generated when a pulse encounters the canopy and the last return when it encounters bare earth. In practice, there are many pulses that only provide one reflection; therefore, the first return can be considered equal to the last return. Modern sensors can record at least three reflections, but often only the first and last are eventually utilized.

4.2.2 Components of ALS

A typical airborne laser scanner can be divided into three components; the laser ranging unit, the opto-mechanical scanner, and the control and processing unit (Wehr and Lohr, 1999). The ranging unit is comprised of the emitting laser and the electro-optical receiver. The transmitting and receiving apertures (typically 8–15cm diameter) are mounted so that the transmitting and receiving paths share the same optical path. This assures that object surface points, illuminated by the laser, are always in the Field Of View (FOV) of the optical receiver. The narrow divergence of the laser beam defines the Instantaneous Field Of View (IFOV). Typically, the IFOV ranges from 0.3 mrad to 2 mrad. The theoretical physical limit of the IFOV is determined by the diffraction of light, which causes image blurring. Therefore, the IFOV is a function of the transmitting aperture D and the wavelength of the laser light l. For spatially coherent light, the diffraction-limited IFOV is given by:

$$\text{IFOCdiff} = \frac{2.44 \frac{\lambda}{D}}{}$$

(Equation 4.1)

The IFOV of the receiving optics must not be smaller than that of the transmitted laser beam. Due to the very narrow IFOV of the laser, the optical beam has to be moved across the flight direction, in order to obtain the area coverage required for surveying. The second dimension is realised by the forward motion of the airplane. Thus, laser scanning means the deflecting of a ranging beam in a certain pattern, so that an object's surface is sampled with a high-point density. A typical airborne laser scanner system is shown in Figure 4.1.

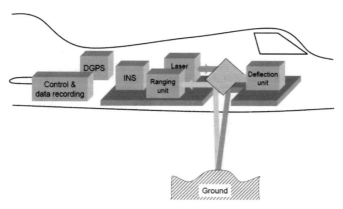

Figure 4.1: Typical airborne laser scanner system (Brenner et al., 2006)

There are many LiDAR systems available on the market. Figure 4.2 shows examples of scanners and their characteristics.

There are two types of lasers; the first is pulsed (time of flight ranging) and the second is Continuous Wave (CW – side tone ranging). Figure 4.3 shows these laser types. The difference between them is their range, range resolution, and range accuracy. CW ranging, resolution, and accuracy, can be improved by using higher modulation frequencies. A pulse laser with high power is also available.

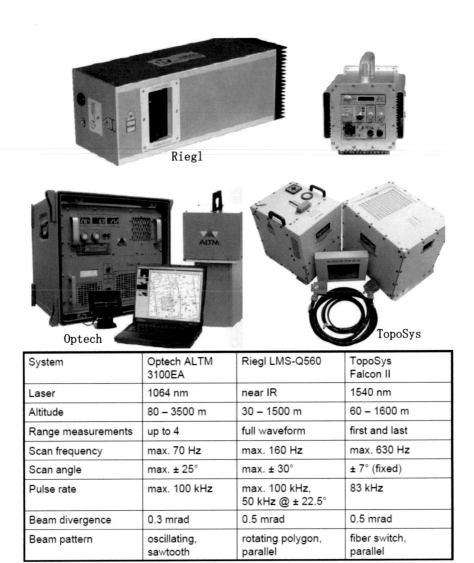

System	Optech ALTM 3100EA	Riegl LMS-Q560	TopoSys Falcon II
Laser	1064 nm	near IR	1540 nm
Altitude	80 – 3500 m	30 – 1500 m	60 – 1600 m
Range measurements	up to 4	full waveform	first and last
Scan frequency	max. 70 Hz	max. 160 Hz	max. 630 Hz
Scan angle	max. ± 25°	max. ± 30°	± 7° (fixed)
Pulse rate	max. 100 kHz	max. 100 kHz, 50 kHz @ ± 22.5°	83 kHz
Beam divergence	0.3 mrad	0.5 mrad	0.5 mrad
Beam pattern	oscillating, sawtooth	rotating polygon, parallel	fiber switch, parallel

Figure 4.2: Example of scanners and their characteristics (Brenner et al., 2006)

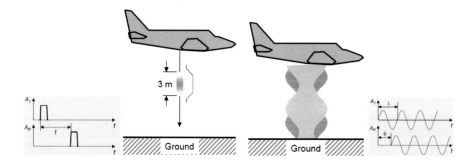

Figure 4.3: Pulsed laser operation (a) and continuous laser operation (b) (Brenner et al., 2006)

A LiDAR system utilises four techniques for swatch scanning. These are oscillating, rotating polygon, mutating mirror, and fibre switch, as shown in Figure 4.4.

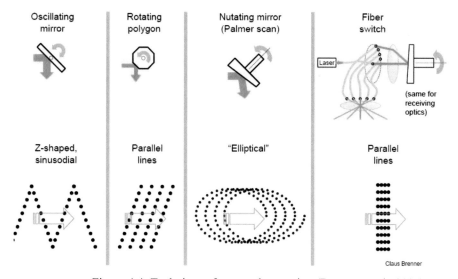

Figure 4.4: Techniques for swatch scanning (Brenner et al., 2006)

4.2.3 Technical parameters

In ALS, technical parameters can be divided into point measurement, scan characteristic, point characteristic, and point cloud. The main technical parameters of ALS can be found in Baltsavias (1999).

Point measurement:
The return times of an emitted laser pulse from the surfaces of a landscape are used to measure the distance from the emission point on an airborne platform to the landscape. By electronically analysing the waveform of a returned pulse, the round-trip time of the pulse can be measured. Therefore, the range from the point of emission to the landscape can be obtained by multiplying the speed of light with half of the return time. An array of range measurements, which is typically linear, is called a scan. Because the range and the position of pulse emissions, and the attitude of the line of sight are known, the position of points on the landscape can be displayed in a 3D frame. Ranges within a scan are measured at rates of 5 kHz and above. The newest systems are able to measure at rates of approximately 100 kHz.

Scan characteristics:
The measurement rate, the scan angle, the flying height, and the aircraft speed, will determined the spacing between points. The scan angle usually ranges from 1° to 65°. Flying heights normally range from 100m to 1000m, but some new and more advanced systems, can be used at heights up to 3km. Therefore, point spacing can range anywhere from 0.1m to 5m.

Point characteristics:
The signal strength of the reflected pulse can be recorded because objects on the ground differ in material composition and height. There are several reflections of a pulse that may be detected. The first reflected pulse is assumed to contain more hits from vegetation, than the second pulse. Therefore, first pulse returns are usually used in orthophoto production, whilst second pulse returns, are used for bare earth measurement applications. Most systems also use it to measure the strength (also known called as the intensity) of the returned pulse. The materials on the landscape have different spectral characteristics, and because of this, a low resolution image of the landscape can be obtained from the strength of the returned pulse. Typically, the radiation used in LiDAR is in the IR part of the EM spectrum. Therefore, materials such as vegetation will tend to appear bright, earth and asphalt will appear dark, and deep-water bodies will absorb radiation. Because of this, reflectance can be used to some extent, for classification. Some ALS systems also capture imagery during scanning. Therefore, an RGB triplet can also be associated with each point.

Point cloud:
When all of the scans are aggregated, a cloud of points in a 3D reference frame is obtained. ALS point clouds are usually large and dense; often containing millions of points.

4.2.4 Data intensity

LiDAR intensity values are a measure of the return signal strength. They are relative rather than absolute, and vary with altitude, atmospheric conditions, bi-directional reflectance properties, and the reflectivity of the target. In terms of laser scanning returns, it is possible to have multiple echoes due to obstructions. Reflectance varies with material characteristics as well as the light used, and different materials have different reflectance (Jeong HS et al., 2002). Consequently, intensities may give useful information in identifying certain objects in a LiDAR point cloud. From previous literature, 'asphalt' (often called bitumen in Asia) has an intensity value of 10~20%, 'grass' approximately 50%, 'tree' 30~60%, and 'house roof' of 20~30%. Because each class has a different intensity, separation between objects can be established. Figure 4.5 shows the reflectivity of various wave lengths of Infra-red in LiDAR.

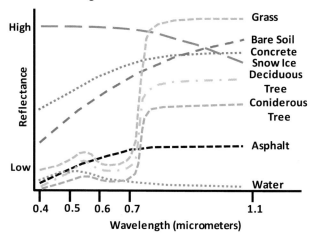

Figure 4.5: Reflectivity of various wave lengths of infra-red in LiDAR (Jensen, 1989)

4.3   Issues concerning ALS data

4.3.1 Data file format

Recently, LiDAR point clouds were stored in two main formats, namely ASCII text and LAS files. ASCII files are formatted with the attributes or fields of the LiDAR points separated by commas, which are also described as Comma-Separated-Values, or a CSV file. The second competing format is LAS; a standardized binary format for storing LiDAR points. Because these two types of data format are usually huge in size, in much of the industry LiDAR points are normally converted to raster. However, the conversion to raster yields significant drawbacks that the user must be willing to accept, such as the loss of attribution, dataset richness, and precision, that results from resampling a dataset.

4.3.2 Vertical and horizontal accuracy

Maksimovic et al. (1999), presented data requirements for advanced modelling of flooding in urban areas. The presentation emphasized the need for high resolution data in a terrain model, for the modelling of urban flooding. The author stated that a 'virtual reservoir', which is an advanced modelling technique, is really important to keep and exchange surcharged water between pipe network and surface flooding. To make a real breakthrough and to model the dynamics of flooding in urban areas reliably, the author expressed the need for high resolution data in a terrain model, which is rarely available in the required form, and at an affordable cost. The presentation deals with data requirements, possible means of gathering the data, some problems of handling large quantities of data, data pre-processing problems, and the application of GIS in matching DTM (Digital Terrain Models) with runoff simulation modules.

Haile and Rienjes (2005) mentioned in their paper that representation topography of rivers and floodplains is a really important aspect for hydraulic flood modelling. A low resolution DTM is normally useful for rural areas, however, in urban areas, the DTM is not suitable as many features, such as road, dykes, buildings, and river banks, will affect flow dynamics and flood propagation. Therefore, high resolution DTM is needed in the model setup to replicate the real topography in order to improve simulation results. The paper discusses the effects of LiDAR DTM resolution in flood modelling in detail. DTM resolution from 2.5m to 15m with an interval of 2.5m has been studied. The conclusion from the study is that DTM resolution has a significant effect on simulation results. The flood simulation characteristics that are affected are inundation extent (up to 15000 m2), flow velocity, flood depth (more than 100%), and flow patterns, across the model domain. The overall conclusion is that the accurate simulation of topography has a significant effect on flood simulation results.

Schumann et al., (2007) compare DTM to LiDAR, topographic contours, and SRTM with 1D hydrodynamic HEC-RAS model, to produce information about water stages during flood events. The different DTM data was validated using reference elevation data distributed across the low-lying flood prone area. The accuracy of each DTM, expressed by RMSEDTM is given by:

$$RSME_{DEM} = \frac{\sqrt{\sum_{i=1}^{n} (E_{Ri} - E_{DEMi})^2}}{n}$$

(Equation4.2)

where,

E_R - reference elevation data

E_{DTM} - elevation data provided by different DTMs

n - total number of reference data used

In terms of data accuracy, as expected, LiDAR data is the best with RMSELiDAR (0.23m), followed by RMSECoutour (0.95m), and finally RMSESRTM (1.55m). However, there are two limitations of LiDAR data, namely:

i.   It only provides discrete surface height samples and not continuous coverage
ii. It's availability is very much limited by economic constrains

The authors also evaluate remote sensing based waterline modelling and again, the results were as expected, with LiDAR data as the best with RMSELiDAR (0.35m), followed by RMSECoutour (0.7m), and finally RMSESRTM (1.07m). The conclusion from this research is that LiDAR data is at present, the most reliable source to reproduce topographic data. This paper shows that LiDAR data is the best source of DTM, compared with other sources, such as contour interpolation and SRTM. The limitation (i) of LiDAR data, gives the opportunity to study the filtering algorithm in detail.

4.3.3 Systematic errors

The overall error budget of an airborne LiDAR system has been the subject of numerous publications. As shown in Figure 4.6, several potential error sources that exist are:

i.   Errors related to the LiDAR instrument itself, which includes the laser rangefinder and scanner subsystems, and errors caused by the finite size of the laser footprint on the ground

ii. Errors determined by the Position and Orientation System (POS), which is required for geo-positioning data collection and processing

iii. Errors in POS/LiDAR interrelations - since the LiDAR data (range and angle measurements) are blended with the POS data, the errors associated with POS/ LiDAR interrelations can play a significant role in the error budget

iv. Errors introduced during data handling and processing, including interpolation and filtering errors, various human errors, and others. During data handling and processing, the overall error budget may be reduced if advanced calibration and optimization or "smoothing" algorithms are applied to the dataset.

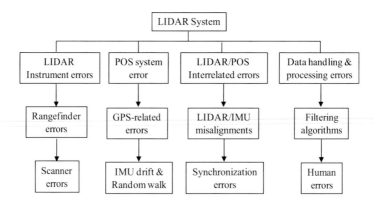

Figure 4.6: Sources of potential errors

4.4 Processing of LiDAR data

4.4.1 Modelling of systematic errors

Error modelling is based on an assessment of the deficiencies of a real system and differences between an ideal and real environment. The error model permits us to predict the errors of the laser points. Comparing simulated errors with observed errors in real datasets, will allow plausible explanations if the errors are modelled on the basis of system imperfections. A most desirable side-effect is that suitable explanations increase the confidence of users in laser altimetry products. The modelling of systematic errors, according to physical causes, has obvious limitations as it is impossible to link remaining errors to a particular system deficiency. Once this limit is reached, the model can be further refined by adding additional parameters that have no direct physical or geometrical meaning.

4.4.2 Filtering and classification

LiDAR collects data from the first surface with which laser beams interact. The DTM generation needs to identify the terrain points on the bare earth, and to remove non-terrain point hits on vegetation, buildings, and other constructions above the ground. Removing unwanted measurements, or finding a ground surface from a mixture of ground and vegetation measurements, is referred to as filtering and classification. Researchers have developed many filtering and classification algorithms. However, they face difficulties in practice, due to the variety and complexity of objects in urban environments, interacting with the discontinuous features of the bare earth.

Existing filtering algorithms can be roughly grouped into four classes. The first class is similar to mathematical morphological filtering. Vosselman (2000) proposed a filtering method, which is closely related to the erosion operator used in mathematical morphology. Height differences between two nearby points are used to determine an optimal filtering function, in order to preserve terrain features. Sithole (2001) and Roggero (2001) varied the method developed by Vosselman. Sithole used a cone-shaped local operator, whose slope was altered to the slope of the terrain, in order to improve the performance of the algorithm in steep terrains. Roggero's filter estimates the local slope in a local linear regression criterion.

The second class of filters works progressively. Part of the bare earth points are first identified and then used to construct an initial Triangulated Irregular Network (TIN). More and more terrain points are identified based on this TIN and then added to classify even more points. Axelsson (1999) derives a TIN from neighborhood minima, and then progressively makes the laser point cloud more dense. This process is called densification. In each iteration a point is added to the TIN, provided the point meets certain criteria in relation to the triangle that contains it. The criteria are that the angle a point makes to the plane of the triangle must be below a certain threshold, and a point must be within a minimum distance of the nearest triangle node. At the end of each iteration, the TIN and the data-derived thresholds are recomputed. The iterative process ends when no more points are below the threshold. Sohn and Dowman (2002) used a two-step process to progressively make a TIN more dense i.e., Downward Densification and Upward Densification. The first is used to obtain an initial description of the bare earth, whilst the second refines the initial TIN. The final TIN is a representation of the bare earth, and the points not included in the TIN are treated as objects.

The third class of algorithms progressively makes the DTM more dense in order to approximate the bare earth. Elmqvist (2002) estimated the ground surface by employing active shape models. A deformable model is used to fit the bare earth by means of energy minimization progressively. Hu (2003), Wack and Wimmer (2002), and Pfeifer & Stadler (2001) used a hierarchical approach, which is similar to the image pyramid method. A coarse DTM is generated at the top level first. and the DTM is then refined hierarchically.

The last class of filters is based on segments. Points are segmented by clustering analysis, region growth, or edge detection techniques, based on height, normal, curvature, slope, or gradient differences, within a small neighbourhood. These segments are then classified based on their contextual information (Sithole and Vosselman, 2004).

4.4.3 Thinning

Density reduction or a 'thinning' algorithm is a useful tool when the volume of a dataset is so large and/or complex that it prevents or inhibits analysis. The tool is often necessary when attempting to portray data at a smaller scale. Thinning the dataset down to a more manageable volume can be done by eliminating those points with negligible new information content. The data can then be expressed as a subset of the original set. If done properly, the data reduction will be both accurate and effective: accurate in that the new set captures the meaning of the original set, and effective in that it is an easier version with which to work (Brady and Ford, 1990).

A point whose removal does not change the topology of the image, of which it is a part, is referred to as a 'simple point'. Points that do impact topology are sometimes called 'mass points'. The concept of simple points is therefore, essential to image transformations (such as density reduction algorithms), which strive to preserve topological features. Bertrand suggests that in thinning an image, one should delete simple points, which are not end points. The simple point condition, prevents an undesired change in the topology of the image, while the non-end point criterion, saves useful information pertaining to the shape of the object (Bertrand, 1994 and 1995).The problem in applying an approximation technique to surface modelling is how to sample only the significant data points from a set of surface points. Ideally, highly curved areas should be sampled densely, whilst less curved areas are sampled more sparsely. This process is important for object recognition in computer vision and to surface modelling at various scales (Li, 1993).

Generalization is considered to have been performed well when the natural topography of an area may be optimally recovered from the limited information portrayed on the map. Cartographic generalization may involve the selection of objects to be omitted, the simplification of lines or features, the combination or summarization of objects, and/or the displacement of objects. Of course, some of these options may not be suitable for specific data types. In continuous data of equal types, the 'information trend' is conveyed to the map by the low frequencies in the data, while the majority of the information content is contained in the higher frequencies.

Chapter 5

FILTERING ALGORITHM

5.1 Introduction

Even though LiDAR has increased our ability to obtain high resolution data, the processing of such data is still a challenging job. LiDAR data processing tasks include the "modelling of systematic errors," "filtering," "feature detection," and "thinning." Of these tasks, filtering and quality control, pose the greatest challenges, consuming an estimated 60 to 80% of the total processing time, which underlines the necessity for on-going research in this area (Schumann et al., 2007).

LiDAR collects data from the first surface with which the laser beams interact. The generation of a DTM involves the identification of bare earth points, and the removal of non-terrain points, such as vegetation, buildings, and other objects. The process of removing points, or finding a ground surface from a mixture of ground and vegetation measurements, is referred to as filtering and classification. To date, a number of filtering and classification algorithms have been developed. Some of these algorithms have been published, while others have unknown details due to proprietary restrictions. The following sections describe some of the existing filter algorithms and the improvement of one of these algorithms for better urban flood modelling.

5.2 Discussion about common features amongst the filter algorithms

The alternative ways of conceiving objects and bare earth adopted by different algorithms can differ slightly. This difference may lead to different filter concepts and approaches. In order to understand the filters, especially from a design perspective, it is worthwhile studying the common features of these filter algorithms. The information gathered can then be used in the design of an improved algorithm. Filter algorithms are built and developed from a combination of different elements. The discussion here focuses on identifying the essential characteristics of these filters.

5.2.1 Data Structure

Typical output data from a LiDAR survey is an irregular 3D point cloud. Some filter algorithms used to filter the raw point cloud into a desired end product already exist (Axelsson, 1999; Kraus and Pfeifer, 2001; Sithole, 2005). However, some other filter algorithms (Brovelli, 2004;

Elmqvist, 2002) resample the raw point cloud into a regular grid before filtering, in order to take advantage of image processing techniques.

5.2.2 Filtering a neighbourhood

There are three possible ways of filtering a neighbourhood.

i. *Point-to-Point (1:1)* - In this method, two points are compared at a time. This is based on the relative position of the two points. If the output of the function is above a certain threshold, then it will be assumed to belong to an object.

ii. *Point-to-Points (1:m)* - In this method, the neighbouring points surrounding the point of interest, are used to resolve a function. The point of interest is classified based on the output of the function. Only one point is classified at a time.

iii. *Points-to-Points (n:m)* - In this method, several points are used to resolve the function. The point of interest is classified based on the output of the function. More than one point is classified at a time.

5.2.3 Classification of points

In some filter algorithms, the classification of points is done in a single step, while others classify points in multiple steps, using iteration. The advantage of a single step algorithm is computational speed, but the accuracy is less than that of the algorithm using iteration. During each iteration, more information about the neighbourhood of a point is gathered and thus a much more reliable classification can be obtained.

5.2.4 Process for filtered data

There are two types of process that can be performed on the filtered data. One of them is called culling, which is usually found in algorithms that operate on irregularly spaced point clouds. In culling, a filtered point is removed from a point cloud. The other process is called replacement, which is usually found in algorithms that operate on regularly spaced point clouds. In a replacement, a filtered point is not removed, but it is returned to the point cloud with a different height; which is usually interpolated from its neighbourhood.

5.2.5 Using external data

External information, such as satellite images, vector maps, and land use maps, can be used to augment the filtering process. Most filters rely solely on the information contained in the LiDAR, but some other filters, particularly those that aim to detect buildings, attempt to use other information to enhance the filtering process. At times, not many filters make use of external data.

5.3 Filtering concepts

Every filter makes an assumption about the structure of bare earth points in a local neighbourhood. For example, bare earth points in a neighbourhood must fit a certain given threshold. In general terms, most existing filter algorithms can be categorised in four groups.

5.3.1 Morphological Filtering

The first group of algorithms is based on mathematical morphological filtering. Vosselman (2000) proposed such filtering, which is closely related to the erosion operator used in mathematical morphology. The height difference between two adjacent points is used to determine the optimal filtering function that preserves the terrain features.

5.3.2 Progressive Filtering

The second group can be described as progressive filters. In these filters, some of the bare earth points are identified first and then used to construct an initial Triangulated Irregular Network (TIN). More terrain points are identified based on this TIN and are added to classify further points (Axelsson, 1999). A sparse TIN is derived from neighbourhood minima, and then progressively made denser. In each iteration, a point is contrasted with the TIN, and if it meets a certain criterion it is added to the TIN. At the end of an iteration, the TIN and the criterion are recomputed. The iterative process ends when no further points conform to the criterion.

5.3.3 Active shape Filtering

The third group of algorithms includes those that progressively increase the density of the points for the DTM in order to approximate the bare earth (e.g., Elmqvist, 2002). The ground surface is determined by employing an active shape model. A deformable model is used to fit the bare earth by progressively minimising the energy associated with the active shape model. When applied to the LiDAR data, the active shape model behaves like a membrane, floating up from underneath the data points. The manner in which the membrane sticks to the data points is determined by an energy function. For the membrane to stick to the ground points, it has to be chosen in such a way, that the energy function is minimized. Hu (2003), Wack and Wimmer (2002), and Pfeifer and Stadler (2001) used a hierarchical approach, which is similar to the pyramid method used by Adelson et al., (1984). In this approach, a coarse DTM is generated at the top level first and then refined hierarchically.

5.3.4 Clustering / segmentation

The fourth group of filter algorithms is based on segments. Points are segmented using cluster analysis, region growth, or edge detection techniques, based on height, normal curvature, slope, or gradient differences, within a small neighbourhood. The segments are then classified according to their contextual information (e.g., Sithole and Vosselman, 2004).

5.4 Research concerning the improvement of filtering algorithms

In research by Klemen Zakse and Norbert Pfeifer (2004) on an improved morphological filter for selecting relief points from a LiDAR point cloud in steep areas with dense vegetation, they decided to improve the morphologic filter. They improved its efficiency in steep areas with dense vegetation by using data from the first echo. A regularly distributed grid approach was used in order to define an appropriate trend surface, because an accurate global trend is analytically too hard to set (at least in the case of rough relief). Therefore, a first order trend surface (plane) is generated within every grid cell, which should be approximately as large as an average canopy of the tree. This value is a lower limit and can be used if the point density is higher than one point per tree canopy. One might expect problems on the border between the grid areas, because local trend surfaces do not merge into a continuous surface; however, these gaps are very small with dense and regularly distributed data. A local trend surface was defined in two steps, which are:

    i.   The first echo data is used in the first step, and

    ii.   The trend is readjusted after analysis of the residuals of the first step trend surface.

The last echo points that pass the first step are used for a trend surface computation. The points from the closest neighbouring areas are also included in the trend determination, because many vegetation points are still present after the first step filtering. Therefore, using neighbouring areas provides a more generalized trend, which is more probable. It is also more probable that the points below the trend are ground points than the points above. Therefore, all of the points that are below the threshold, which is defined by distribution parameters (derived from standard deviation of the residuals to the trend surface), are accepted as ground points. Moreover, if the standard deviation is small enough (regarding the accuracy of the measurements), all of the points are accepted.

Xuelian Meng (2005), in his research, proposed a new slope and elevation based filter (OTFL) to extract the ground points from airborne LiDAR data. This filter applies a multi-directional scanning technique to label the ground points based on the slope change and elevation difference to the nearest labelled ground points. This filter also considers the local elevation difference to prevent accumulated errors. The identified ground points, in the former scanning directions, remain in the label of the next scanning direction; allowing the filter to utilize the former results and increase the chances of finding a nearer ground point. This research compares the performance of the OTFL filter with the one-dimensional and bi-directional scanning (OBL) filter. The quantitative error analysis based on the smaller study site, shows that the OTFL filter improved accuracy from 94.7891% to 97.3969%, and the Kappa values increased from 0.8918 to 0.9463. The OTFL filter can apply a smaller and suitable maximum elevation threshold, which is critical when applied to high-relief areas. These experiments prove that the OBL filter is vulnerable to the selection of scanning directions and the distribution of objects, which are however, improved by the OTFL filter. The OBL filter fails with larger areas on small hills, whereas the OTFL filter generates better results without obvious errors, according to the observations of the field survey.

Marc Bartels and Hong Wei (2006) in their research on the segmentation of LiDAR data using measures of distribution, present an unsupervised *skewness balancing* segmentation algorithm, to separate object and ground points efficiently from high resolution LiDAR point clouds, by exploiting measures of distribution. Their proposed algorithm works on balancing the distribution of points in the LiDAR data. Statistical measures of distribution are independent of the relative position of the points. That is why they do not have to be regularly arranged in a DSM. Therefore, the proposed technique works on both gridded data and point clouds. As kurtosis and skewness both express the characteristics of the point cloud distribution, they can be treated equally as termination criteria in a segmentation algorithm. In this unsupervised segmentation algorithm, skewness is chosen as a measure to describe the point cloud distribution. Thus, this algorithm is called *skewness balancing* and works as follows: First, the skewness of the point cloud is calculated; if it is greater than zero, peaks dominate the point cloud distribution. Thus, the highest value of the point cloud is removed by classifying it as an object point. In order to separate all ground and object points, these steps are executed iteratively whilst the skewness of the point cloud is greater than zero. The remaining points in the point cloud finally belong to the ground. The results presented by Bartels amd Wei have shown that the proposed algorithm is robust and has the potential for commercial applications.

Jong-Suk Yoon et al., (2006) in their research on the extraction of bare ground points from airborne LiDAR data in mountainous forested areas, propose a filtering method to separate bare ground points from multiple return laser point clouds over mountainous forested areas in Korea. The process of filtering in this study consists of three parts; the initial separation, and a first and second separation. The processing starts with raw LiDAR data, which is stored in an ASCII format of the first and last returns. Because of existing trees, points are distributed mainly for the canopy and ground. Among the first and last returns, some points are backscattered from the same location – these are called 'singular returns'. Since the Singular Returns (SRs) are stored in the first and last return data files simultaneously, the separation of the SRs attempts to bring efficiency to the subsequent processing. The results of the initial separation are Pure First

Returns (PFR), Singular Returns (SR), and Pure Last Returns (PLR). Their proposed algorithm has been further verified using International Society for Photogrammetry and Remote Sensing (ISPRS) reference data.

Xuelian Meng et al., (2008) conducted research into so-called morphology-based building detection from airborne LiDAR data. This research presented a morphological building detection method to identify buildings by gradually removing non-building pixels. First, a ground filtering algorithm separates the ground from buildings, trees, and other objects. Then, an analytical approach further removes the remaining non-building pixels, using size, shape, height, building element structure, and height difference, between the first and last returns. The results show that this method provides a comparative performance with an overall accuracy of 95.46%, as in the Austin urban area study site.

C. K. Wang and Y. H. Tseng (2010) are concerned about how to generate a DTM from airborne LiDAR data. They propose an adaptive dual-directional filter based on an existing slope filter to construct the DTM. The main difference between them is the designed filter shape. The dual-directional adaptive filter is designed to be use along an alternative direction, in one dimension while in original slope filter it is applied to be use in its whole adjacent covered window. The different directions adopted in the dual-directional adaptive filter can avoid an over-filtering situation. The ISPRS test data has been used for the evaluation of this filter. The variance, omission errors, and commission errors, are shown for comparison. From the test, it is seen that he dual-directional adaptive filter method shows better results in handling data for abrupt surfaces compared to the original slope-based filter and the commercial software TerraScan. It is also shows that the method used in dual-directional adaptive filter has better performance in avoiding an over-filtering situation and can maintain good accuracy compared with other methods.

Chapter 6

EVALUATION OF CURRENT FILTERING ALGORITHMS

6.1 Existing filtering algorithms

Seven different filtering algorithms are considered from the different groups referred to above. They are: two morphological algorithms (first group), a TIN based algorithm (second group), and four shape-based algorithms (third group). These seven algorithms are the Progressive morphological filter 1D/2D, Elevation Threshold with Expand Window (ETEW) Filter, Maximum local slope filter, Iterative polynomial fitting filter, and Polynomial 2-surface filter. These are the algorithms that are most widely used by researchers, as well as industry practitioners.

6.1.1 Progressive morphological filter 1D/2D

The main assumption of this algorithm is that points in a given height range adjacent to a particular measurement point are for bare earth. In the 1D Progressive Morphological algorithm the lowest point in a neighbourhood is labelled as a terrain point. By gradually increasing the window size and using elevation difference thresholds, it removes measurements for different sized non-ground objects, while preserving ground data. The maximum elevation difference threshold can be set either to a fixed value to ensure the removal of large and low buildings in an urban area, or to the largest elevation difference in a particular area. The filtering window can be a one-dimensional line or a two-dimensional rectangle, or any other shape. When a line window is used, the opening operation is applied to both x and y directions at each step, to ensure that the non-ground objects are removed (Zhang et al., 2003).

The two fundamental operations in this algorithm are dilation and erosion. These operations are commonly used to enlarge or reduce the size of features. These concepts can also be extended to the analysis of a continuous surface, such as a digital surface model, as measured by LiDAR data. For a LiDAR measurement $p(x,y,z)$ the dilation of elevation z at x and y is defined as:

$$d_p = \max_{(x_p,y_p)\in w} (z_p)$$

(Equation 6.1)

where d_p is the dilation of elevation, p is the point cloud, x_p is the x coordinate of the point, y_p is the y coordinate of the point, z_p is the height value of the point, and w is the window.

Erosion is the counterpart of dilation, and is defined as:

$$e_p = \min_{(x_p, y_p) \in w} (z_p)$$

(Equation 6.2)

where e_p is the erosion of elevation, P is the point cloud, x_p is the x coordinate of the point, y_p is the y coordinate of the point, z_p is the height value of the point, and W is the window.

The combination of erosion and dilation generates opening and closing operations that can be used to filter the LiDAR data. An erosion of the data set, followed by a dilation, is performed to generate the opening operation, while the closing operation is accomplished by carrying out dilation first and then erosion. An erosion operation can remove tree objects of sizes smaller than the window size. Dilation can be used to restore the shapes of large building objects. The ability of an opening operation to preserve features larger than the window size is very useful in some applications (see also Zhang et al., 2003). For example, for a dense urban area, the measurements of large buildings can be preserved if the morphological filters are applied to the LiDAR measurements. The schematic of this algorithm is shown in Figure 6.1.

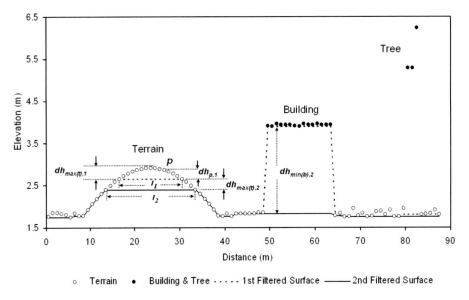

Figure 6.1: Schematic of the Morphological Filter for separation between ground and non-ground measurements (Zhang et al., (2003))

The 2D Progressive Morphological algorithm adopts the same concept. The only difference is that the filtering window is a two-dimensional rectangle or any other shape. Apart from using the

68

line, it also uses a square window, which can perform erosion in the x direction first followed by the y direction. The same rule can be applied to dilation. The weakness of this filter is that the results are influenced by the final window size and threshold value for which the points are expected to be terrain points. Too small a window leads to large building points being labelled as ground points. Too high a threshold leads to many vegetation points being labelled as ground points. The strength of this filter is that the entire process is carried out by gradually increasing the window size, which can be controlled by the user.

This algorithm has the ability to remove non-ground objects, such as buildings and trees, using typical processes including opening, closing, dilatation, and erosion, based on kernel operators. When applied to urban flood modelling, this research has shown that the algorithm has good filtering capabilities of unwanted objects, such as vegetation and cars. However, the same algorithm can also filter out all buildings, which cannot be regarded as a positive because the buildings are needed for flood modelling.

6.1.2 Elevation Threshold with Expand Window (ETEW) Filter

The main assumption of this algorithm is that the elevation differences of neighbouring measurement points are distinct between ground, trees, and buildings, in a limited sized area within a search window (Zhang and Whitman, 2005). Elevation differences in a certain area are used by the algorithm to separate ground and non-ground LiDAR measurements. The non-ground points are identified and removed by the elevation threshold method using an expanding search window. The weakness of this filter is that it is unable to find the correct elevation threshold value. Its strength is that the inherent concept and calculation is rather simple and straightforward.

The ETEW filter can sometimes create abrupt elevation changes of ground measurements near to cell boundaries, because minimum elevation thresholds are different for each cell. Usually, the elevation threshold value will cause the buildings and vegetation points to be labelled as ground points and thus to be included in the DTM. This situation allows the filter to preserve the buildings and vegetation in a complex urban environment. Since the buildings and vegetation represent important features in an urban area, their preservation (without proper handling) may have a considerable impact on the floodwater flow. Furthermore, in relation to urban flood modelling, this algorithm has not performed well in that it produces far too much noise in the resulting DTM. Even though the buildings are not filtered out, as is the case with the previous algorithm, the buildings can only be represented as solid objects.

6.1.3 Maximum local slope filter

The main assumption of this algorithm is that the terrain slopes are different from the slope that can be found between the ground and the top of a building. In this algorithm, a comparison of the local slopes between the LiDAR point and its neighbourhood is used to identify point measurements. These slopes are used to separate the ground from non-ground points. It is assumed that along the boundaries of the ground and non-ground areas, the slopes between the ground and its neighbouring non-ground points are much larger than those between the ground and its neighbouring ground points. This filter uses the slope of the line between any two points in a point set as the criterion for classifying ground points. The technique relies on the premise that the gradient of the terrain's natural slope is distinctly different from the slopes of non-terrain objects (trees, buildings, etc.). Any feature in the laser data that has slopes with gradients larger than a certain predefined threshold is classified as a point that does not belong to the natural terrain's surface (Sithole and Vosselman, 2002). The weakness of this filter is that it is vulnerable to slope change. When applied to steep slopes, the filter usually fails to extract the ground points. The strength of this filter is that it can detect high structures, such as tall buildings, which are often found in many urban environments.

This filter preserves a better shape of ground objects as it is sensitive to small, sharp changes in area, such as shrubs and short walls, which are difficult for most filters. This is because a point is classified as a ground point if the maximum slope of the vectors connecting the point to all of its defined neighbours does not exceed the maximum slope within the study area, which, in an urban environment, may be very large due to the high elevation of buildings. As a result, the final DTM will have too many unwanted objects, such as trees, cars, etc.

6.1.4 Iterative polynomial fitting filter

The assumption of this algorithm is that the lowest point in a set of neighbouring measurement points belongs to the ground. In this algorithm, LiDAR points are classified by selecting ground measurements iteratively from the original data set. The lowest point within a large moving window (which is usually larger than the non-ground object in a particular area) consists of an initial set of ground measurements. For example, ground measurements at the top of a small mountain can be missed because the interpolated surface is too low due to the lack of previously identified ground points at the mountain top (Zhang and Cui, 2007). This error can be recovered by comparing the elevation difference of a candidate ground point to the current surface. When compared with current surface the points at the top of a mountain will be identified as ground measurements, because their elevation differences from the current surface are less than the predefined threshold. However, other non-ground points might be included by mistake. To remove these errors, the fitness of the previous and current surfaces to the ground measurements within a surface interpolation window needs to be introduced as another criterion. If the fitness of the current surface is better than the previous one then the missed ground point is recovered. The weakness of this filter is that its result tends to have a lower value than the real data. Overall, this algorithm has the tendency to produce misleading information and, if the final DTM is directly used for modelling of urban floodplains, (without post-processing) it will certainly have negative consequences.

6.1.5 Polynomial 2-surface filter

The main assumption of this algorithm is that the ground is essentially represented as being continuous, or at least, a piecewise continuous surface. A polynomial 2-surface filter uses a mathematical function to approximate the ground. A least squares adjustment is used to detect the non-ground points as if they were incorrect, by reducing their weights in each iteration calculation (Zhang and Cui, 2007). This method usually requires flat areas and does not work well in mountainous terrains or where the buildings appear more like hills. With this method, only small objects and the boundaries of larger elements can be eliminated, because the edges of very high buildings cannot be detected, due to resolution failure and thresholds. This is because points within a certain vertical distance above the surface are treated as ground points. If used in urban flood modelling work, high buildings and elevated structures, such as elevated roads and flyovers that are included, will create artificial obstacles to the flow. This algorithm also has a tendency to produce too much noise in the resulting DTM, which makes it inappropriate for urban flood modelling applications.

6.1.6 Adaptive TIN filter

The main assumption of this algorithm is that nearby points have similar attributes, while distant points have dissimilar attributes. In this algorithm, the distance of points on the surface of a TIN is used to select ground points from a LiDAR data set. A sparse TIN is derived from neighbourhood minima, and then progressively made denser, depending on the laser data. During iteration, a point is added to the TIN if the point meets certain criteria in relation to the triangle that contains it. The criteria are that the angle a point makes to the triangle must be below a certain threshold and a point must be within the minimum distance of the nearest triangle node. At the end of the iteration, the TIN and data-derived thresholds are recomputed. New thresholds are computed based on the median values estimated from the histograms of an iteration. Histograms are derived for the angle points, to generate TIN facets and the distance to the facet nodes. The iterative process ends when there are no points that are below the threshold (Axelsson, 1999). The weakness of this algorithm is that it requires large data storage and significant computational time for execution. This algorithm can successfully remove small buildings and most bridges, but fails to remove some larger buildings. This is because the algorithm uses a fixed window size to calculate mean values and therefore, fails to remove those objects that are larger than the window size. If the resulting DTM is used for urban floodplain modelling, then the results are likely to be erroneous in those areas where such buildings exist.

6.2   Assessment of current filtering algorithms

The assessment plays an important role in both ground filtering applications and algorithm development. The capability of current filtering algorithms that have been discussed above has been assessed using both quantitative and qualitative assessment. Quantitative assessment was a challenge for LiDAR ground filtering, due to the lack of ground truth data. For this reason, this research has utilized visual inspection and random sampling of ground filtered data in its quantitative assessment, and also for the qualitative assessment.

6.2.1 Quantitative assessment

The quantitative assessment was done in two ways. The first was by comparing the height of objects that had been filtered through the algorithms with the actual measurements. Equation 6.3 was used to evaluate the filter's accuracy. In this assessment, three objects were selected to be tested, namely the divider, the bridge, and the LRT train line.

$$RMSE = \sqrt{\frac{1}{N}\sum_{i=1}^{N}(y_i - x_i)^2}$$

(Equation 6.3)

where y_i = height from filters

x_i = observed height

Table 6.1: Summary of quantitative assessment. The values in this table were obtained by computing the standard statistical measure, RMSE.

Filter Objects	Poly2surf	Slope	ETEW	ATIN	Poly	Morph	Morph2D
Divider	0.292	0.291	0.291	0.264	0.270	0.273	0.294
Light rail train	1.711	1.740	1.694	0.157	0.169	0.153	0.157
Bridges	0.331	0.324	0.324	0.321	1.109	1.028	0.986

It can be seen from Table 6.1, that all of the algorithms can detect dividers quite well. The heights of the detected dividers are very close to the real heights. This is because the characteristics of a divider itself are very close to the bare earth. ATIN, Poly, Morph, and Morph2D, seem to have the capability of detecting the light rail train with a very low RMSE error. As for bridges, almost all of the filters gave low RMSE values, which meant that the difference between the measured height and the filter detected height was similar.

There are Type I and Type II errors in the LiDAR information extraction. A Type I error is the rejection of bare earth points where bare earth is misclassified as objects. A Type II error is the acceptance of object points as bare earth where objects are misclassified as bare earth. This type of error is used to determine the potential influence of the filtering algorithms on the resulting DTM, based on the predominant features in the data set. Referring to this procedure, another way of deriving a quantitative assessment is by calculating the percentage of Type I and Type II errors. Due to the lack of ground truth data, the available ground survey data was combined with the Advanced Spaceborne Thermal Emission and Reflection Radiometer (ASTER) Global

Digital Elevation Model Version 2 (NASA, 2011) which was assumed to be the estimated ground truth for the study area. The ASTER DEM was generated using stereo-pair images collected by the ASTER instrument onboard Terra. In this research, the analysis of the Type I and Type II errors is done by considering the combination of the available ground survey data and ASTER DEM as the ground truth reference.

From the analysis (see Table 6.2), it can be seen that in the overall performance, Type I errors (see Figure 6.2) are more than Type II (see Figure 6.3) errors for all filters except for Poly2surf. This is because most of the algorithms are designed for determining, as far as possible, bare earth points from objects.

Table 6.2: The percentage of Type I and Type II errors calculated from the current filters.

FILTER	TYPE I (%)	TYPE II (%)
ATIN	10.091	1.013
ETEW	14.001	3.396
MORPH2D	10.711	0.024
MORPH	10.369	0.008
POLY2SUR	7.313	24.689
POLY	7.352	2.618
SLOPE	24.158	8.444

Morph and Morph2D seem to have a very low percentage of Type II errors, which suggests that these two filters can generate a DTM quite well. Poly2surf has the highest percentage of Type II errors, but in contrast it has the least Type I errors, which suggests the reason why most object points were preserved in this filter. Slope has the highest number of Type I errors, but also created some Type II errors. Slope works well for both flat and oblique surfaces. However, areas near terraced fields and cliffs may result in unreliable estimations of slope. Classification errors will occur in this kind of area, where some points near the break line (boundary between objects and bare earth) would usually be eliminated. This situation is called over-filtering and the missing points will result in a smoother DTM than the true DTM.

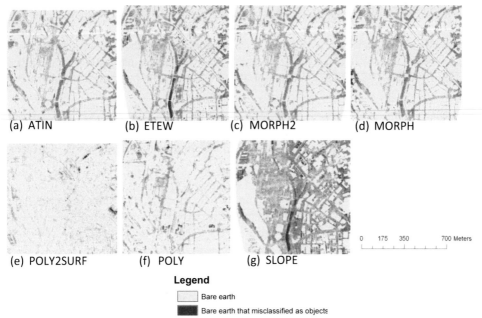

(a) ATIN (b) ETEW (c) MORPH2 (d) MORPH

(e) POLY2SURF (f) POLY (g) SLOPE

0   175   350   700 Meters

Legend

☐ Bare earth

■ Bare earth that misclassified as objects

Figure 6.2: Incorrectly classified points. Red represents the points that were incorrectly classified as objects (Type I errors).

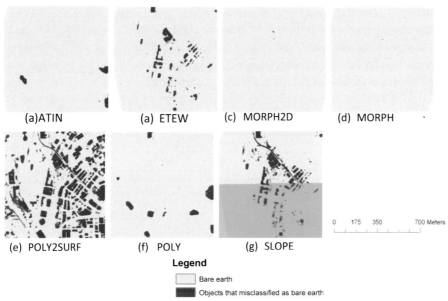

(a)ATIN (a) ETEW (c) MORPH2D (d) MORPH

(e) POLY2SURF (f) POLY (g) SLOPE

0   175   350   700 Meters

Legend

☐ Bare earth

■ Objects that misclassified as bare earth

Figure 6.3: Incorrectly classified points. Red represents the points that were incorrectly classified as bare earth (Type II errors).

6.2.2 Qualitative assessment

A qualitative assessment was undertaken to evaluate all of the algorithms mentioned above. In this assessment, criteria were used that focused on the removal of buildings, flyovers, and bridges, and the capture of curbs and river alignment. The assessment was done by visually assessing performance and giving a mark with a weighted value. This weighted value was based on the filter performance in removing and capturing features. If more than 75% of the features are removed or captured, the filter was given 1 mark; if 50 to 75% of the features are removed or captured, the filter was given 2 marks; if 25 to 50% of the features are removed or captured, the filter was given 3 marks, and if less than 25% of the features are removed or captured, the filter was given 4 marks. The lowest total mark will therefore suggest which filter performs the best, based on the selected criteria. The results show that each filter acts differently and has different features depending on its filtering concept. The evaluation summary is shown in Table 1. It can be seen from the table that the Morph algorithm performs well in removing bridges, buildings, vegetation, and flyovers. Since the Morph2D algorithm uses a similar concept to the Morph algorithm, its results are unsurprisingly close: the only difference is observed with respect to the handling of bridges.

In an urban environment, the difference between the ground points and the top of buildings are so big that some of the filters perform ineffectively. Because the ETW filter is not iterative, the determination of the threshold in a big difference of heights leads to buildings and flyovers being captured, while fewer points are removed, as shown in Figure 4. In the Slope filter, a big difference in heights produces a steep slope; and leads to a large gradient being predefined as the threshold. This situation leads to most of the points being preserved, with a minimal removal of points (see Figure 6.4). With Poly2surf, most of the points are classed as ground points. Therefore, this algorithm performs well in capturing buildings and curbs, but is less efficient in removing bridges, flyovers, and vegetation.

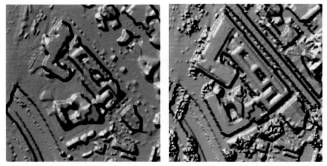

Figure 6.4: Example DTMs generated by ETEW (left) and SLOPE (right), show that most object points have been preserved and the removal of points is minimized, caused by different reasons provided by the filters.

The Polynomial filter results show a good capability of removing features. This is because most of the elevated feature points are missed, due to the lowered interpolated surface, which resulted from the lack of previously identified ground points in the elevated area. As for the ATIN algorithm, most objects smaller than the window size, including buildings and flyovers, are removed successfully, but most of the bridges located near bare earth were not removed. (See Figure 6.5)

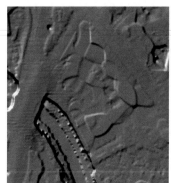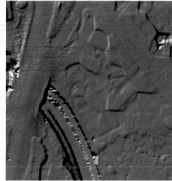

Figure 6.5: Example DTMs generated from Poly (left) and ATIN (right), showing how the filters removed features based on different assumptions.

All filters gave acceptable results for curbs and river alignments (Table 6.3). Morph appears to have the least overall total mark. This suggests a better performance in developing the urban DTM that is required as input for the urban flood model. Supported by this qualitative assessment result, the Morph filter is regarded as the basis for modifying LiDAR data in order to deduce the urban DTM needed for flood modelling.

Table 6.3: Summary of the qualitative assessment

Indicator	Morph	Morph2D	ETEW	Slope	Polynomial	Poly2surf	ATIN
Building Removal	2	2	4	4	4	3	3
Bridge Removal	2	4	3	4	3	4	3
Flyover Removal	2	1	4	3	1	3	2
Curbs Capturing	2	2	1	2	2	2	3
River Alignment	2	3	3	2	3	2	2
Vegetation Removal	2	2	3	3	1	4	2
Total Weighted Value	12	14	18	18	14	18	15

Weighted Value: 1=Excellent, 2=Fair, 3=Acceptable, 4=Poor,

6.3 The weaknesses of current filtering algorithms

The weaknesses of current filtering algorithms commonly come from the two main sources of landscape and data properties.

6.3.1 Landscape properties

In every LiDAR point cloud, some obstacles come from the arrangement and the nature of objects with the bare earth in the landscape. These obstacles can be categorized into four groups:

Numbers of objects
Any filter usually works efficiently in a condition where there are more bare earth points than objects. In urban environment there are usually more objects with bigger sizes, and this condition means that more points are categorized as object points than bare earth. This presents a weakness for current filters because most of them operate on a local neighbourhood basis. These filters will face problems if there are more object points than bare earth points.

Location of objects compared to the bare earth
Generally, the closer an object's location is to the bare earth means the harder it becomes to separate it from the bare earth. There are many objects in an urban environment, such as dividers, embankments, raised platforms, and terraces, which are close to the bare earth and will raise problems for the current filters.

Complication in an object's shape
In an urban environment, although most buildings are regular in shape there are many that are more complicated in shape. Current filters will face a problem, because they operate on small neighbourhoods, where locally object-to-object relations and bare earth-to-object relations are clearly distinguished.

Complication of landscape
Besides complicated object shapes, the complexity of the landscape also appears, especially when it comes to discontinuities. Again, current filters will face a problem, because they usually operate in small neighbourhoods.

6.3.2 Data characteristics

Besides object properties, data properties can also contribute difficulties to filters. However, different to the difficulties presented by landscape properties, difficulties in filtering due to data properties can be controlled to a certain extent. Additionally, these difficulties will not occur in every data set. The data properties that can throw in difficulties to filters include:

The resolution of a point cloud

In a low resolution point cloud, discontinuities become difficult to discern, while in a high resolution point cloud there are numerous noises present. This property has a direct effect on the spatial definition of objects, where less defined objects will be more similar to the bare earth.

Scanning system and strip overlaps

In the scanning system used, the scan angle, the scan swath, and the flying speed of the aircraft, contribute to the variation of point density in strip overlaps. A point cloud is created by combining strips of scans. The resulting point clouds will have higher densities in areas where strips overlap. Because proximity measures in filters depend on point density, in a point cloud formed by combining strips, selective points based on proximity will give difficulties. A proximity threshold used in an area that overlaps may lead to classification of bare earth points as objects (Type I errors) in non-overlapping areas. Correspondingly, proximity thresholds that are used in the non-overlapping areas may lead to classification of object points as bare earth (Type II errors) in overlapping areas.

Lack of bare earth information

Typically, the penetration rate of the LiDAR survey into vegetated areas is lower, at around 25%, than in any other areas (Lindenberger, 1991), especially when scanning is done in paddy areas where the paddy plants are so closed to each other. However, this penetration rate can sometimes be very low because there are very few bare earth samples. With the absence of bare earth samples, it is not possible to achieve meaningful filtering.

6.4   The weakness of current assumptions

Current algorithms distinguish between objects and the bare earth using different assumptions. These assumptions are created based mainly on the designer's concept of the condition for bare earth and also the designer's wish for a simple but useful algorithm. Most current filters operate such that they are based on small neighbourhoods, which mostly depend on locally bare earth to bare earth relations and bare earth to object relations. This chapter discusses the development of new assumptions, and for this purpose the common assumptions in current filters are listed here:

Assumption 1
Points in a given height range within a neighbouring measurement point are for bare earth

Assumption 2
The elevation changes of neighbouring measurement points are distinct between ground, trees, and buildings in a limited sized area within a search window

Assumption 3

The terrain slopes are different and considerably smaller from the slope that can be found between the ground and the top of a building

Assumption 4

The lowest point in a set of neighbouring measurement points belongs to the ground surface

Assumption 5

The ground surface is essentially represented as being continuous or at least as a piecewise continuous surface

Assumption 6

Nearby points have similar attributes while distant points have dissimilar attributes

Depending on the type of landscape for which the filters were designed, all of the listed assumptions are correct, which suggests that there is actually no such thing as a wrong filter, but rather a filter that is applied incorrectly. The current assumption weaknesses mostly come from landscape characteristics. Referring to the number of objects, globally there is more bare earth than objects. However, locally this is not always the case. This gives problems for Assumption 1 through to Assumption 6, because it is possible to have more object points than bare earth points in a neighbourhood. The relation between bare earth and objects also gives problems to Assumption 1 through to Assumption 6, because the relation is in direct proportion to the vertical and lateral separations between them. Referring to this, the closer an object is to the bare earth, the more difficult it becomes to separate it from the bare earth. When objects and the bare earth come nearer to each other, all of the stated assumptions will become invalid.

The previously discussed assessment shows that the current assumptions used are inadequate, especially for urban landscapes. This suggests that there is an insufficiency in the algorithm and hence, shows an inadequacy in the assumptions. This also suggests that better assumptions can be formulated.

6.5 Formulating new assumptions

As we go through the list from Assumption 1 to Assumption 6, a single method considered in every assumption becomes an issue, and the emphasis is placed on working with a combination of several methods rather than one individual method.

The formulation of new assumptions focuses more on the urban landscape and also the consideration to use external data, such as LiDAR intensity and vector data.

Assumption 1

The height range in the neighbourhood of a particular measurement point alone is not enough to perform efficient ground filtering.

Assumption 2

The combination of height range and terrain slopes gives better performance in distinguishing between objects and bare earth

Assumption 3

Buildings are highly elevated objects in an urban area

Assumption 4

Most buildings in an urban area have smoothly sloping roofs

Assumption 5

Buildings should create a closed polygon in vector form. Closed and clean vector polygons are the polygons that represent buildings

Assumption 6

Points that do not create vector closed polygons represent high land and vegetation (see Figure 6.6).

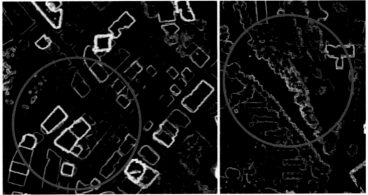

Figure 6.6: Closed and clean polygons that represent buildings (left) and points that do not create closed polygons (right), which usually represent high land and vegetation.

Assumption 7

Based on the condition of the study area, buildings with a certain height threshold will be considered as buildings with basements.

Assumption 8
Buildings with a height of less than 4 metres will be considered as solid objects.

Assumption 9
Based on the condition of the study area, buildings with a height between the height of a building with a basement and a solid object will be considered as a passage building, which can allow flood water to pass through to a certain extent. These points are represented in the DTM as Manning roughness 20.

Assumption 10
The use of LiDAR intensity information helps to determine certain objects that can easily be distinguished from bare earth

Assumption 11
Curbs lay very close to the bare earth

Assumption 12
Curbs usually create a continuous line

From the review and the Assumptions 1 to 6 from the selected current filters, it is found that each selected current filter applies an individual method for its filtering processing. From the assessments it is found that there are some disadvantages to only considering one method in the filtering process. To overcome this, Assumptions 1 and 2 are formulated so that the two methods, which are the difference in height and the calculation of slope, are combined to do the filter processing. In this step, the separation between objects and bare earth is done generally and does not focus on any special features.

Going through Assumptions 1 to 6 from the selected current filters, it is clearly seen that the objective for each filter is only to distinguish between objects and bare earth, and not to try and differentiate features from the object points. The new Assumptions 3 through to Assumption 9 are formulated to differentiate buildings from the object points. This is because buildings have been identified as one of the essential objects in an urban environment that have a direct effect on the flooding pattern (see Chapter 3). These new assumptions are also formulated, not only to differentiate the buildings, but as a basis of representing them in a DTM. Besides buildings, elevated roads are also considered to be essential objects. Assumption 10 is formulated to separate elevated roads from the object points. Before this process can be carried out, Assumption 3 and Assumption 4, which were formulated earlier, are adopted and used to select all elevated objects with a smooth surface, including elevated roads. Then the information from the value of intensity captured from LiDAR is used to detect asphalt, which is the main component in road construction, in order to classify the elevated road associated with this object with other smooth surface objects.

Assumption 11 and Assumption 12 are formulated especially for distinguishing curbs from the object points. Assuming that curbs lay very close to bare earth, points with a height 5-7cm higher than the average bare earth are selected. From this group of points, a vectorization is done, and all continuous lines are considered as curbs, while others are ignored.

6.6   Selection of a suitable filtering algorithm

From the review, Morph appears to have the lowest overall total mark. This suggests a better performance in developing an urban DTM that is required as input for the urban flood model. Supported by this qualitative assessment result, the Morph filter is regarded as the basis for modifying LiDAR data in order to deduce the urban DTM needed for flood modelling. This is mainly due to its strength in separating ground and non-ground points, by increasing the window size iteratively. In particular, this algorithm gives the possibility that the user can be in control while removing and preserving objects.

Chapter 7

Development of Modified Progressive Morphological Algorithm (MPMA)

7.1   Introduction

As indicated in Chapter 6, out of the seven algorithms reviewed the Progressive Morphological Algorithm (PMA) has been shown to be more promising than the other algorithms and as such it was selected for further development. This is mainly due to its strength in separating ground and non-ground points by increasing the window size iteratively. In particular, the algorithm gives the possibility that the user can be in control while removing and preserving objects. The code of this improved algorithm was written in Visual Basic.

7.2   Modification of filtering algorithm framework

The framework of the work done to improve the algorithm is focused on the following issues:

    i.   To detect buildings and to classify them into: solid buildings, passage buildings and buildings with basements;

    ii.   To remove simple bridges along a river;

    iii.   To detect elevated roads and light train lines, to remove related objects and to reconstruct structures that are underneath;

    iv.   To retain curbs

    v.   To detect riverbanks and to interpolate points between the banks where the river network is modelled with a 1D model (MIKE11), and to generate a DTM for use with a 2D model (MIKE21);

    vi.   To test the usefulness of the algorithm by carrying out the 1D/2D modelling work for a study area.

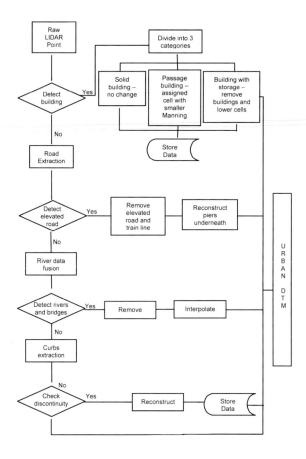

Figure 7.1: Flowchart for the development of the new filtering algorithm

In order to fulfill the filtering algorithm objectives, the development process is divided into three major parts. Each part is concerned with one of the selected essential objects in an urban environment, namely buildings, bridges, elevated roads and river banks and curbs. The overall flowchart for this new filtering algorithm is shown in Figure 7.1.

7.3   Phase I: Detection and modification of buildings

Detection and classification of buildings was carried out as a three step procedure:
Step 1: Detection of buildings from point cloud data
Step 2: Separation of objects from point cloud data
Step 3: Classification of buildings

84

7.3.1 Detection of buildings from point cloud using slope concept

In this step, the existing morphological algorithm is improved by initially detecting and labeling all the buildings. This process is done by manipulating the slope concept. It is based on the following assumptions:

i.   Terrain slopes are different from the slopes that are found between the ground and the top of a building;
ii. Buildings are highly elevated objects in an urban area;
iii. Most buildings in an urban area have smoothly sloped roofs;

Initially, points are labelled as "High" or "Low" based on their elevation. In the present development, the value for "High" points was left to be user-defined. Then, for each point in the point cloud the slope in percentage rise is calculated for 3 x 3 neighbourhoods around every point (Figure 7.2 – right). The maximum value of the slope from these neighbourhoods is taken as an attribute for that point (see Figure 7.2 – left).

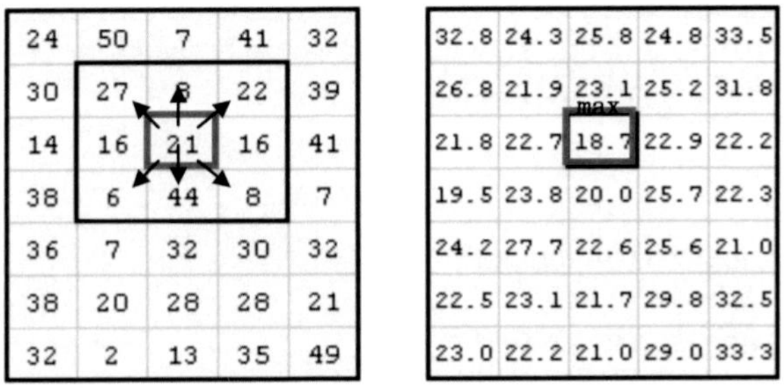

Figure 7.2: The concept of the slope calculation

Based on the slope, points are divided into two classes; "Steep" and "Slight". The initial slope is user-defined and is usually the slope between the building which has a minimum value of height, and the surface. The members of the Steep class are those points representing vegetation and walls of buildings, while the members of the Slight class are the representatives of building roofs and relatively flat areas on the surface. Points that represent a building wall are points with a "High and Steep" label (See Figure 7.3). These points are selected and converted to a vector form as polygons. Based on the assumption that a building should lead to the creation of a closed polygon in vector form, all created polygons are tested to see if they are closed or not. Points that do not create closed polygons are usually represented as high land and vegetation. These unclosed polygons are then removed. The closed polygons are converted back to raster form as a grid.

85

0 62.5 125   250 Meters

Figure 7.3: Steep points are in brighter colour while Slight points are the in darker colour

The conceptual step for the classification between surface and building is presented as follow.

Data: Object point, o
Begin

Set threshold for low point, l [where l = maximum height for average surface point]
Set threshold for initial slope, s [where s = slope between the building which have minimum value of height and the surface]

While V= objects;
            For each v > l do
            | Label v as high point
            Else
            | Label v as low point
            End

            Calculate Slope
            Slope = Rise *100
                        Run
            These rates are computed for a 3*3 neighborhood around every point.

            Set threshold for slope, s
            For each v > s do
            | Label v as steep [points representing high-rise vegetations and walls
            | of buildings]

            Else
            | Label v as slight [points representing building roofs and relatively flat
            | areas on the surface]
            End

86

Identified building

For each v = high point and steep
| Assign v as o [object]
| Convert to vector
Else
| Assign v as s[surface]
End

For each o in vector form
| For o = Closed polygon
| | Label = b[building]
| | Convert to raster (DTM)
| Else
| | Removed
| End
End

End
End

7.3.2 Separation of other unwanted objects from point clouds

The purpose of this procedure is to remove the vegetation and other objects from point cloud data. The procedure is based on the original Progressive Morphological algorithm. Apart from the advantage for the user in controlling the separation of ground and non-ground points, this algorithm is also found to be useful because of its relatively simple structure as well as its effectiveness in separating ground and non-ground points. The conceptual step for the classification between surface and building is presented as follows.

Data: Point cloud, V
Begin

Set Cell size, c
Set Maximum window size, max
Set Terrain slope, s
Set initial elevation difference threshold, dh0
Set maximum elevation difference, dhMAX

Determine the minimum and maximum of x and y values
Determine the number of rows(m) and column(n)
Create 2D array for LiDAR points LiDAR[m,n]
Determine series of window size, ws where ws<=max

Set dhi = dh0

For each ws do
    For i = 1 to m
        Pi = LiDAR[i]
        Assign value Z(erosion)
        Assign value Zf(dilation)

        For j = 1 to n
            If Z[j] – Zf[j] > dhi then
                flag[i,j] =ws
        End
    End
End

For i = 1 to m
    For j = 1 to n
        If LiDAR[i,j](x)>0andLiDAR[i,j](y) > 0
            If flag[i,j] =0
                LiDAR[I,j] = bare earth (be)
            Else
                LiDAR[I,j] = object (o)
    End
End

For LiDAR[i,j] = object (o)
    If points = building
        preserved
    Else
        removed
    End
End
End

7.3.3 Categorization of building objects

In this step, the original morphological filtering algorithm was improved by assigning the buildings that have been preserved with the three types of properties, namely, buildings with a basement, passage buildings and solid objects (See Figure 7.4). A site survey should be done in the study area to determine the characteristic for these three types of properties. Data need to be collected including the height of the building, numbers of buildings with basement, numbers of passage buildings and numbers of solid objects. When the data has been collected, the percentages of buildings that represent those three types of properties are calculated and the buildings are categorized based on their group height.

Figure 7.4: Example of a building with basement (left), a solid object (top right corner) and a passage building (bottom right corner).

7.3.4 Attached building with basement with downward expansion

For points detected as buildings with a basement, the point elevation is changed by lowering it to a certain depth below the ground surface depending on the number of basement levels. In this case, a place for the retention of flood water is created (See Figure 7.5). The determination of depth is dependent on the characteristics of the study area. The examples of such characteristics are the average height of ground surface, the elevation of the highest building and the basement height. In this algorithm, these characteristic are defined by the user based on the study area. If no information is input for these characteristics, standard values are used.

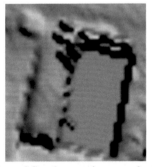

Figure 7.5: A place for the retention of flood water is created when the buildings are categorized as buildings with basement.

7.3.5 Determined Manning value for passage building

Points with a passage building label have an elevation value that is similar to the average ground surface. This point is then assigned a Manning value (roughness parameter) of 20 (n=0.05) while the rest is assigned a Manning value of 30 (n=0.033) and exported as an ASCII file. Smaller Manning values lead to the flow having some resistance when passing through the passage building pixel. This ASCII file is used to generate the dfs2 file (grid file in MIKE21) which is used later in flood modelling. For solid building points, the original elevation is kept. When this process is done, points are merged back with the pre-DTM points.

The setting of the Manning's roughness coefficients for flood plains follows the standard settings for MIKE 21. A Manning number (M) of 30 (n=0.033) was found to be a practical starting value in most floodplain applications. By altering Cowan's (1956) procedure which was developed for estimating n values for channels, the following equation can be used to estimate n values for a flood plain (Arcement, 1984):

$$n=(nb +n1 +n2 +n3 +n4)m \qquad\qquad\qquad \text{(Equation 7. 1)}$$

where nb is a base value of n for the flood plain's natural bare soil surface, n1 is a correction factor for the effect of surface irregularities on the flood plain, n2 is a value for variations in shape and size of the flood-plain cross section, assumed to equal 0.0, n3 is a value for obstructions on the flood plain, n4 is a value for vegetation on the flood plain and m is a correction factor for sinuosity of the flood plain, equal to 1.0 in this study.

The values for nb, n1, n2, n3, n4 and m can be determined from Table 2.1 (Refer Chapter 2).

The conceptual process for the building categorization is presented as follows:

Data: Building point, B
Begin
      Set Building with basement height, bbh
      Set Basement height, bh
      Set Passage building height, pbh

      While building found do
         For each B > (ground +bbh) do
          Label B as building with basement
             For each B=building with basement
                Lowering the point value by (ground – bh)
           End
      End

For each (ground+bbh)>B>(ground+pbh) do
    Label B as passage building
        For each B= passage building
            Assign point value to (ground)
            Give Manning value lower than the overall cell
            Create dfs2 grids
        End
    End

For each B < 4 do
    Label B as solid building
End
    End
End

7.4 Phase II: Detection and modification of elevated road, bridges and river banks

7.4.1 Detection and removal of elevated roads and train lines

In reference to the detection of buildings, points were labeled as 'High' or 'Low' and 'Steep' or 'Slight' based on their height and associated slope, respectively. The value for 'High' points is user defined and it is usually the maximum value of the average surface (bare earth). Values lower than the value of 'High' are considered as 'Low'. Members of the 'Steep' class are those points which represent vegetation and walls of buildings, while members of the 'Slight' class are representatives of relatively flat surface areas which are based on the following assumption:

  i.   Terrain slopes are usually different from the slopes occurring between the ground and the top of an object (elevated road/train)

  ii.   Elevated road/train are highly elevated objects in a scene

The combination of point with 'High' and 'Slight' labels are points that represent relatively flat areas on high surfaces, including elevated roads, elevated train lines, and building roofs, as shown in Figure 7.6. This concept is used as the basis of detecting elevated roads and trains lines. The next step is to separate the elevated road/train lines from other objects. The separation process is performed by selecting the points that have intensity values between acceptable ranges for the type of road material being detected (in this case asphalt/bitumen), and for this the intensity feature of LiDAR is used. As mentioned above the intensity value of 'asphalt' is approximately 10~20%. By using its intensity value, roads can be identified from the points cloud.

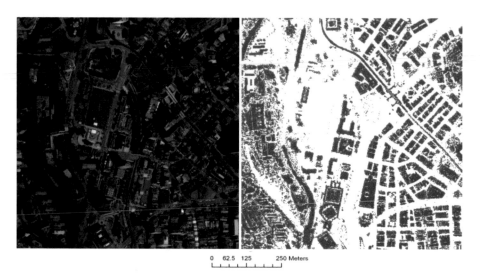

0   62.5 125   250 Meters

Figure 7.6: Points with 'High' and 'Slight' labels, which indicated relatively flat areas on high surfaces.

In order to see the correlation between intensity values and road objects, 2908 of sample points from several areas that represent roads in the points cloud were extracted. These sample points were checked for their intensity values. From this operation it was found that even though there were some errors, the majority of the sample points (81%) show the values in the range between 10% and 20%, which corresponds to the range of intensities of asphalt values (Table 7.1). Figure 7.7 illustrates the relation between point cloud data and the data that have intensity values of roads. No values above 30% were recorded as the sample given in Figure 3 concerns only the road area.

Table 7.1: Observation of the sample points of roads

TYPE	NUMBER OF POINTS	PERCENTAGE
INTENSITY (10%-20%)	2369	81.46
OTHERS	539	18.54

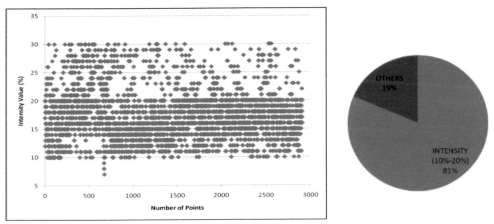

Figure 7.7: Scatter graph and pie chart show the relation between point clouds in road feature with its intensity value.

Even though the intensity values returned by the scanning unit were noisy, sections of road material were typically uniform for road/rail lines and as such they can be distinguished relatively easily. By searching for a particular intensity range it is possible to extract the points that refer to road/rail lines on the elevated surface. Figure 7.8 shows the points that correspond to elevated road/train lines after the separation process.

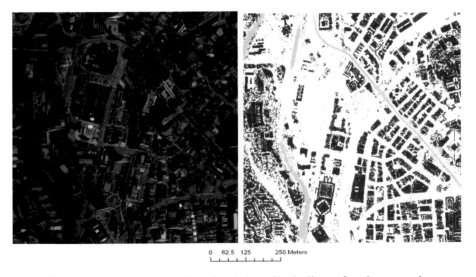

Figure 7.8: Points that correspond to elevated road/train lines after the separation process. The green points show the selected candidate points for elevated road/train while the red points show all other points.

93

The conceptual design of the separation between elevated roads and other objects presented as follow.

Data: Points from building part, B

Begin

> For each B = "High" and "Slight"
> | For 10<intensity<20
> | | Label B as elevated road
> | Else
> | | Label B as others
> | End
> End
>
> For B = elevated road (E)
> | Remove
> End

End

7.4.2 Options in the incorporation of piles

In this step, the new algorithm gives an option to add the piles underneath the elevated road/train lines, if the information is available. This process is made optional, so if the information about the dimension of piles, the distance between piles, and their location are not available, the filtering and classification process can still be performed.

7.4.3 Incorporation of piles underneath the elevated road and rail lines

The new algorithm allows for incorporation of piles underneath the elevated road/train lines, if the information is available. Within the algorithm, this process is made optional so that if the information about the dimension of piles, the distance between them, and their location, is not available, the filtering and classification process can still be performed.

In this option, points that have been identified as elevated road/train lines are converted into vector form as lines, before they are removed from the DTM. Using these lines as a basis, the pile shapes are incorporated in the vector form, based on locally surveyed information. The information needed for this reconstruction process includes the pile's height, width, length, and the distance between the piles. The algorithm also permits the piles to be placed in the middle or perpendicular position with the line, as shown in Figure 7.9. Once the vector reconstruction process was complete, the created polygons were converted into a grid and then merged back with the DTM.

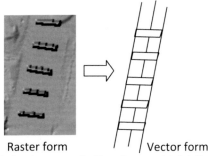

Raster form   Vector form

Figure 7.9: Reconstruction of piles in vector (left) and raster (right) forms.

The conceptual design of this process is presented as follow.

Data: Points from first part, E
Begin
        Set Height, h
        Set Width, w
        Set Length, l
        Set Length between, lb
        Set starting point coordinate x,y

        h = get input from user
        w = get input from user
        l = get input from user
        lb = get input from user
        x= get input from user
        y= get input from user

        For each E
                Convert to Vector as line (vl)
                Construct piles using h,w,l and lb using vl as a basis
                Put the piles in the middle and perpendicular with the line using x,y as basis
                Convert the polygon to grid
        End
End

7.4.4 Removal of bridges across the river and interpolation between river banks

The process of detection and removal of bridges and river surface interpolation has been implemented within the new algorithm by using the data fusion concept. The river polygon data are overlaid over the points cloud data using the buffer of 5m on both sides of the river and the points within the river polygon are then extracted. From the selected points, the points with intensity values that correspond to asphalt (i.e., 10%-20%) are used to identify location of bridges and to remove them from the DTM. Information about the location of bridges is cross-referenced with the geometry of 1D model to ensure that all bridges (and culverts) are correctly incorporated within the 1D model. The river banks are then interpolated between the left and right bank elevation values by applying the Kriging interpolation method and the resulting DTM is used in setting up the 2D model and coupling it with the 1D model.

Data: Points from first part, B
Begin

    Set river polygon with 5m buffer, PB

    Identify river area
        For each B = "completely within"
        | Label B as river area
        Else
        | Label B as others
        End

    Detect bridges
        For B = river area
            Check the intensity value
            For each B = >=10 AND <=20
                Label B as bridges
            Else
                Label B as others
        End
End

7.5 Phase III: Detection and modification of curbs and representation of close-to-earth vegetation

The overall aspects of Phase III are:

    i.   Ability to detected curbs using data fusion and vectorisation processes.
    ii.   Ability to recover the curbs using the sub-grid concept.

iii.   Ability to detect closed-to-earth vegetation, to remove them from a DTM and to incorporate with the appropriate roughness coefficient (Manning's value).

7.5.1 Assumptions for curbs detection

Curb detection on the urban DTM is an important and necessary task for the preparation of a detailed urban surface as an input for urban flood model. Usually the height of the curbs ranges between 5 to 10 cm. At the same time LiDAR data has vertical accuracy of 5 to 15 cm. It is a challenging task to determine the curb height accurately because in many places the curb height is too small or there is no curb at all. The curb height is important in simulation model because of the location of street inlets and the possible entry of the water from the street to private ground. Street inlets are located at the foot of curb wall. In the case where the street inlets are located on the top of curb wall by mistakes, flood modelling is going to be affected.

In this thesis a semiautomatic approach for detection of road is applied. This approach combines the vector data and raw LiDAR point cloud without converting it into any other raster format. The idea behind this approach is that raw points preserve the originality of the surface before they are converted to any other formats. It works based on assumptions that:

i.   Curbs lies very closed to the bare earth
ii.   Curbs usually created continuous line

First, the street network is collected from vector maps. Regarding this network as defining the location of the curbs, curb points are extracted from the LiDAR point cloud by searching through the points on both sides of the road. The point searching region is limited by the input of the road boundary lines from road polygon maps with a 30cm buffer for both sides. Within the searching region, points with a difference of between 5 to 10cm from the average road height are selected as candidates for curb lines. The average road height is provided as input to this algorithm. The curb extraction algorithm is guided by points for the road polygon and searches for the next curb points. Subsequently it extracts points bounded by the searching region. Figure 7.10 shows the searching regions developed from vector maps.

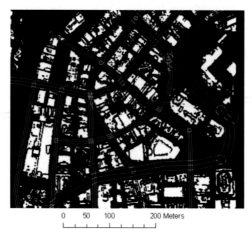

Figure 7.10: The searching regions developed from vector maps

7.5.2 Converting curbs to vector form

Curb candidate points extracted from the point cloud are converted into a grid and then converted into a vector map (See Figure 7.11). The process of converting them into a vector map takes into account the orientation of groups of cells and fits lines to them. This is based on the sweep line Voronoi algorithm devised by Steven Fortune (1986) along with some filtering techniques. All cells that have a neighbouring cell with a value sufficiently different to it are turned into vectors.

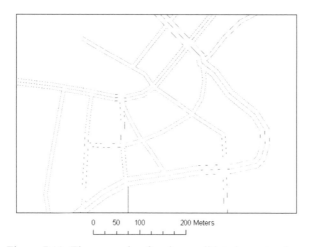

Figure 7.11: The example of curbs candidate in vector form

7.5.3 Recovering of curbs using sub-grid

A high resolution grid is often needed to divide the data set into small tiles geographically. This may result in a poor performance when dealing with large study area. To address this issue, a sub-grid was adopted to create a finer set of grid tiles in certain important areas and a coarser grid set for quick representation of large areas. A grid merging process was proposed to produce seamless landscape scenes consisting of multiple tiles for different grid layers. First, the original DTM data needs to be processed to generate a DTM that represents the whole study area. Then the finest grid is generated where the curbs have been identified through the vectorization process using the ArcGIS10 syntax function:

Con(IsNull(Finer_grid),[Coarser_grid],[Finer_grid])

The workflow of the developed terrain using sub-grid concept is shown in Figure 7.12. The idea is to set threshold resolutions for different levels. When the level of detail is coarser, the threshold should be larger, while when the detail needed is such as to generate curbs the threshold should be smaller. When visualization starts, the coarser grid set is used initially to prevent processing an unnecessarily detailed terrain model.

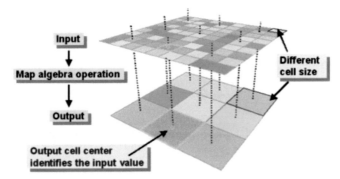

Figure 7.12: The workflow of the developed terrain using sub-grid concept

The conceptual design of this process is presented as follow.

Data: Points cloud, B
Begin

       Set road polygon with buffer, R

       Identify curbs in searching region, R
             For each B > 5 and B< 10
             | Label B as curbs, C
             Else
             | Label B as others, O
             End

       Vectorization of curbs
             For each C
             | Convert to vector, CP
             End

       Recovery of curbs
             For each CP
             | Recovered curbs using sub-grid
             End
End

7.5.4 Detection of closed-to-earth vegetation and represented it with Manning's roughness coefficient

A general problem in modelling surface runoff in urban areas is the lack of information about the characteristics and distribution of urban soils. Another major disadvantage is the lack of knowledge about the spatial distribution of soil sealing (e.g. grass, bush), which highly influences the amount of surface runoff. In urban hydrology modellers have to cope with the paradoxical situation where there is such a tremendous information gap concerning input data especially for those areas where detailed spatial information would be necessary to make useful predictions for flooding (i.e. urban areas). Whereas information on urban surface is very difficult to obtain and usually associated with time-consuming field work, the degree of surface cover can be estimated using the distribution of Manning's roughness coefficient.

In this research, the close-to-earth vegetation extraction algorithm encapsulates the points with an elevation less than 30cm compared to the average urban surface in the study area and excluding the points that have been detected as curbs. Usually the area covered by this kind of vegetation generates an identifiable area (polygon). Once detected, the corresponding points remain in the DTM but the points are assigned an appropriate Manning's value.

Data: Points cloud, B

Begin

        Identify close-to-earth vegetation in searching region, V

            For each V > 5cm and B< 30cm

            | Label V as curbs, CTE

            Else

            | Label V as others, O

            End

            For each CTE

                Assign point value to (ground)

                Give appropiratie Manning value

                Create dfs2 grids

            End

            Deletion of CTE

                For each CTE

                | Delete CTE

                End

End

Note: dfs2 is a MIKE21 grid file format – (DHI Release Note 2011 http://releasenotes.dhigroup.com/2011/MIKEFLOODrelinf.htm)

In this research, the Manning value adopted for MIKE21 model is 30 (n=0.033) for the entire surface except for the cells that represent passage buildings. The Manning value of 20 (n=0.05) is used for a passage building, to emulate disturbances due to local obstacles (e.g., shops, bike racks, etc.). For an area covered by close-to-earth vegetation, the Manning value of 40 (n=0.025) is used. This value is determined from Table 2.3.

Chapter 8

Case Study – Kuala Lumpur

8.1   Introduction

The proposed filtering algorithm, MPMA was tested on real data in order to evaluate the efficiency of the proposed algorithm to accommodate urban flood models. This chapter examines and discusses the results of the tests. The results were compared by five difference aspects which include the comparison of DTM, flood depth, flood extent, flood velocity and the distribution of the surface roughness.

Three different cases were used in the evaluation.

i.   The first case comes from the dataset used to generate the DTM in Phase I of the MPMA development as described in Chapter 7, which considers the detection and classification of buildings.

ii.   The second case results from Phase II of the MPMA development. The dataset used here is the result of the combination of Phase I and the detection and removal of elevated road and rail lines. Additionally, the corresponding algorithm is used to detect river banks and interpolate within the banks.

iii. The third case comes from the dataset used to generate the DTM resulting from the complete development of MPMA. This contains Phase I, Phase II and Phase III (recovery of curbs and the assignment of appropriate Manning's roughness coefficients to close-to-earth vegetation)

For purposes of the DTM comparison, two dataset have been set-up where one dataset contains the DTM resulting from the existing filtering algorithm PMA while the other includes the results of using the proposed filtering algorithm, MPMA. Both datasets were classified into the classes: bare earth and object, using the respective filtering algorithms. In order to evaluate the influence of the DTM on the results of an urban flood model in terms of flood depth, flood extent, flood velocity and the distribution of the surface roughness, a reference dataset was acquired from the Department of Irrigation and Drainage Malaysia (DID) in which the measurements of flood depth and flood extent were recorded based on actual flood events.

8.1.1 Case Study Area

The Klang River Basin is the most densely populated region in the country with an estimated population of over 3.6 million which is growing at almost 5% per year. The basin encompasses the Federal Territory of Kuala Lumpur and parts of Selangor's districts of Hulu Langat, Gombak, Petaling, Sepang, Klang and Kuala Langat.

The Klang River Basin has been radically modified from its natural forested condition by human activities in the pursuit of urban and industrial development through activities including clearing of forests, cultivation, construction of dams, river canalisation works, drainage of swamps, and construction of roads, railways and other infrastructure. Development in the Klang River basin has been concentrated in the rivers banks and their floodplains in order to gain ready access to water for domestic and industrial purposes. As a result of the rapid land development, these areas are always prone to flooding.

Previously, DHI Water and Environment (2004) developed and calibrated a MIKE 11 model of the overall Klang river basin as part of the Klang River Basin Environmental Improvement and Flood Mitigation Project of the Department of Irrigation and Drainage Malaysia. In their study, DHI modelled the entire Klang river basin. The recent case study area concerns a small part (2km x 2km) of the Klang River basin (See Figure 8.1).

It is located at the confluence of Sg. Gombak and Sg. Klang at the heart of Kuala Lumpur city centre namely at Masjid Jamek which floods frequently. Just upstream of the confluence is the Jalan Tun Perak bridge over Klang river, which is a significant flood monitoring location as there exist a flood gauging station located here. The other significant flood gauging location is downstream of this confluence at the Jalan Sultan Sulaiman Bridge over Klang River. Although the major flood mitigation works within the Kuala Lumpur city have, to a large extent, been implemented, flooding in the city is still frequent and severe. This is causing disruption to various activities in the city as well as extensive flood losses suffered by flood victims.

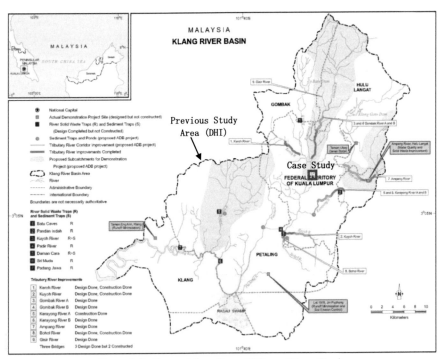

Figure8.1: General location of Klang River Basin

8.2 Problem description

8.2.1 Urban Surface

Urbanisation and industrialization in the river basin have been rapid with major portions of agricultural and ex-mining land being converted for urban use. It is estimated that about 50 percent of the Klang River Basin has been developed for residential, commercial, industrial and institutional use. As a result of the extensive and rapid urban development in the basin area, problems have emerged in the form of river overbank floods, flash floods that afflict clogged drainage systems and river environment degeneration. These problems have prompted the commissioning of a number of additional flood mitigation and river environment enhancement programmes because the severity of the problems and the associated social and economic costs are escalating with further urbanization.

The Klang Basin suffers flooding on a regular basis despite the fact that the government has embarked on the comprehensive flood mitigation program involving expenditure of RM500 million, construction of large dams and extensive channel works. Flood mitigation works has been rendered ineffective because flood discharges have increased threefold since 1986 due to the effect of urbanisation, which has been underestimated in the past. There are many factors that have contributed to the increase in flood discharges including creation of large areas of impervious surface, filling of swamps and filling of flood plain, channelization works and forest clearing. Field experiments carried out by the Department of Irrigation and Drainage in 2001 have shown that a forest catchment in the Klang Valley is capable of absorbing up to 100mm of rainfall in the first hour of a storm while the same soils under urban conditions are typically can absorb only 20mm of rainfall under the same condition.

Whilst structural measures are a key component in reducing the impact of flooding on existing development, the Malaysia Government has also recognized the importance of non-structural measures. These include the implementation of strong planning and building control which needing the strong and reliable support from urban flood models.

Urban environments can contain a vast variety of features (or objects) that have a role in storing and diverting flows during flood events. In this respect, buildings are the most significant objects. In addition to buildings, there are also many small geometric "discontinuities" such as elevated roads, stairs, pavement curbs, fences and other objects which can play an important role in diverting flows that are generated over the urban surface. These features can be undetected by airborne LiDAR. When these features are not adequately represented it is highly likely that the flood model will not be able to produce satisfactory results.

8.2.2 DTM and Urban Flood Model

The Government of Malaysia, through the Department of Irrigation and Drainage (DID) as its implementing agency, now intends to approach the problems more holistically through the integrated river basin management (IRBM) approach to improve the river environment and flood mitigation works in the Klang River Basin.

The role of modelling within urban flood management is in complementing the acquisition of data to improve the information and understanding about the performance of a given drainage network, taking into account the associated urban terrain. Advances in urban flood modelling have been made through the use of 2D hydraulic numerical models in the last decade. These advances offer the potential to predict the local pattern and timing of flood depth and velocity, enabling informed flood risk zoning and improved emergency planning. Considerable attention has been given to the acquisition of good geometric and topographical data at an adequate resolution to describe the primary features of the flow paths through the urban area. Researchers have shown that with the availability of high resolution DTMs derived from airborne LiDAR these models can theoretically be routinely parameterized to represent considerable topographic complexity, even in urban areas where the potential exists to represent flows at the scale of individual buildings. However, computational constraints on the use of fine resolution DTM for 2D urban flood modeling require model discretizations at scale well below those achievable with LiDAR and thus unable to make optimal use of this emerging data stream

In relation to the representation of urban features in the DTM, it was observed from the results of the research that continuous overland flow along roads cannot be represented in the model if the road width is less than $\sqrt{2}$ (Pixel resolution). The effective flow path between buildings also disappears with the increase in pixel size. In terms of pixel resolution on model results, it was found that there is an increase in flow speed with the decrease in pixel resolution within the flood plain. Also, relatively larger flow speed and water depth existed in built-up areas compared to those in non-built up areas. Two reasons have been explored to justify this statement including;

i.   Reduction of the slope of the DTM with the decrease in spatial resolution.

ii.   Loss of flow paths between buildings with the increase of pixel size.

The slope reduction is a consequence of the loss of local undulations in the terrain as the pixel size increases, whereas a significant reduction does not occur along a longitudinal section through the entire terrain.

8.3 Data

There are three types of data usually involved in urban flood modeling which include the geographical data, hydrology and hydraulics data and flood related data.

8.3.1 Geographical data

A digital terrain model (DTM) is one of the most essential items of information that flood managers need in present-day practice. A DTM is essentially a topographic map that includes spot elevations for the terrain and data for its properties. Correspondingly, the term DTM (or digital elevation model), although usually associated with the land surface, refers to the elevations of any surface for any object. In urban flood management, DTMs are needed for an analysis of the terrain topography, for setting up 2D models, processing model results, delineating flood hazards, producing flood maps, estimating damages, and evaluating various mitigation measures. Typically, a DTM data set can be obtained from ground surveys (e.g., total stations together with Global Positioning System – GPS), aerial stereo photography, satellite stereo imagery, airborne laser scanning or by digitising points/contours from an analogue format such as a paper map, so that they can be stored and displayed within a GIS package and then interpolated.

Airborne Laser Scanning (ALS) or Light Detection and Ranging (LiDAR) is one of the most common techniques that is used to measure the elevation of an area accurately and economically in the context of projects involving cost/benefit analysis. It can deliver information on terrain levels to a desired resolution. The end result of an ALS survey is a large number of spot elevations which need careful processing. In this research, the LiDAR System used was Riegl LMS Q560 with Full Waveform Analysis for unlimited target echoes. The Pulse Rate was 75kHz or 75,000 points per second. This system used all echoes and all intensities. The beam size was less than 1.5m diameter with 60 degrees swath width. The data has 40% side lap where the laser distance is approximatly 700m. The flying height was approximately 700m above ground level with the average of 100knots flying Speed. The platform for this system was a helicopter (Bell 206b JetRanger).

8.3.2 Hydrology and hydraulic data

The Study area is on the west coast of Peninsular Malaysia and is generally hot and wet throughout the year without much variation. Nevertheless, the climate can be loosely defined by the following seasons:

i.   The north-east monsoon from December to March
ii.   A transitional period from April to May
iii. The south-west monsoon from June to September; and
iv.   A transitional period from October to November

In addition, it is also characterized by uniform high temperature, high relative humidity, heavy rainfall and little wind. The average annual rainfall depth in the Study area is about 2,400 mm. The monthly variation of the rainfall is given in Figure 8.2. The highest rainfall occurs in the months of April and November with a mean of 280 mm. The lowest rainfall occurs in the month of June with a mean of 115 mm. The wet seasons occur in the transitional periods between the monsoons, from March to April and from October to November. In addition to rains associated with the monsoons, rainstorms due to convection occur occasionally throughout the year during late afternoons. The rainstorms last for a short duration, are isolated and usually of very high intensity.

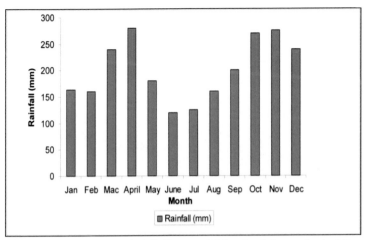

Figure 8.2: Mean Monthly Rainfall (Source: Malaysian Meteorological Services)

The temperature throughout the year is fairly constant with a mean of 27°C. The highest temperature falls at 1pm with an average of 32°C and the lowest temperature falls at 7am with an average of 23°C. As the relative humidity is very closely related to the surrounding temperature, its variation throughout the year is also a minimum with an average value of 82%. The highest and lowest values coincide with the lowest and highest temperatures with values of 96% and 62%, respectively. Another parameter that is also related closely to the temperature and relative humidity is the evaporation rate. The evaporation depths for an open water pan are measured at around 1500 mm per annum and monthly mean of around 125mm/month. The highest evaporation occurs in February to March and the lowest evaporation occurs in November to December, although the difference is marginal.

The wind speed seldom exceeds 10 m/s and is considered calm for around 36% of the time. The duration of sunshine hours per day is fairly constant for the whole year and the mean daily sunshine hours are 6 hours. In the surrounding of study area, several hydrological stations have been installed by DID to collect data. Figure 8.3 and Table 8.1 presents an inventory of the hydrological stations.

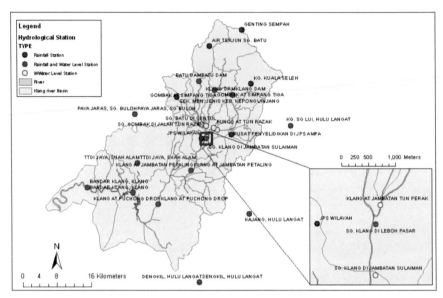

Figure 8.3: Location of rainfall and water level stations around the study area

Table 8.1 Inventory of rainfall and water level stations

STATION NO	STATION NAME	STATION TYPE
3115402	Paya Jaras, Sg. Buloh	Rainfall/Waterlevel
3115401	Ttdi Jaya, Shah Alam	Rainfall/Waterlevel
3015432	Taman Sri Muda, Shah Alam	Rainfall/Waterlevel
3014401	Bandar Klang, Klang	Rainfall/Waterlevel
3118445	Kg. Sg Lui, Hulu Langat	Rainfall/Waterlevel
2917401	Kajang, Hulu Langat	Rainfall/Waterlevel
2816441	Dengkil, Hulu Langat	Rainfall/Waterlevel
3217005	Gombak At Simpang Tiga	Rainfall/Waterlevel
3116432	Sg. Klang Di Leboh Pasar	Rainfall/Waterlevel
3116434	Sg. Batu Di Sentul	Waterlevel
3116430	Sg. Klang Di Jambatan Sulaiman	Waterlevel
3116433	Sg. Gombak Di Jalan Tun Razak	Waterlevel
3117402	Sg. Klang Di Lorong Yap Kwan Seng	Waterlevel
3015001	Klang At Puchong Drop	Rainfall/Waterlevel
3216403	Batu Dam	Rainfall/Waterlevel
3217435	Klang Dam	Rainfall/Waterlevel
3116435	Klang At Jambatan Tun Perak	Waterlevel
3117403	Bunos At Tun Razak	Waterlevel
3106401	Klang At Jambatan Petaling	Waterlevel
3217005	Gombak At Simpang Tiga	Rainfall/Waterlevel

110

3216004	Sek. Men. Jenis Keb. Kepong\Jinjang	Rainfall
3317004	Genting Sempah	Rainfall
3117070	Pusat Penyelidikan Di Jps Ampa	Rainfall
3317001	Air Terjun Sg. Batu	Rainfall
3217004	Kg. Kuala Seleh	Rainfall
3116432	Sg. Klang Di Leboh Pasar	Rainfall/Waterlevel
3015001	Klang At Puchong Drop	Rainfall/Waterlevel
3216403	Batu Dam	Rainfall/Waterlevel
3217435	Klang Dam	Rainfall/Waterlevel
3116004	Jps Wilayah	Rainfall
3115402	Paya Jaras, Sg Buloh	Rainfall/Waterlevel
3115401	Ttdi Jaya, Shah Alam	Rainfall/Waterlevel
3015432	Taman Sri Muda, Shah Alam	Rainfall/Waterlevel
3014401	Bandar Klang, Klang	Rainfall/Waterlevel
3118445	Kg. Sg Lui, Hulu Langat	Rainfall/Waterlevel
2917401	Kajang, Hulu Langat	Rainfall/Waterlevel
2816441	Dengkil, Hulu Langat	Rainfall/Waterlevel
3106401	Klang At Jambatan Petaling	Rainfall/Waterlevel

In this research, there are two significant flood monitoring locations which are located in the upstream and downstream of Sg. Gombak and Sg. Klang confluence. In the upstream of this confluence, the flood gauging station is at Jalan Tun Perak bridge over Klang river and in the downstream of this confluence is at the Jalan Sultan Sulaiman Bridge over Klang River. These two flood monitoring stations are describe as below:

3116435 Klang River at Tun Perak Bridge

Station   3116435 Klang River at Tun Perak bridge

Location   Downstream of the bridge on Jalan Tun Perak

Catchment   122 km2
Area

Equipment   Stilling well attached to vertical wall of channel, with float directly coupled to chart recorder (to be confirmed), analog-digital converter providing input to telemetry system.

Stream gauging from the pedestrian walkway on the bridge, using a winch mounted on a wheeled trolley to cantilever the current meter out over the river.

Features	The recorder is situated about one meter downstream of a sheet pile cut-off wall that extends across the channel at the upstream end of the engineered channel. The top of the sheet pile wall is slightly lower in the centre than at the sides, so that flow over the wall is higher at the centre. A small amount of spill around the end of the sheet piling causes some turbulence near the stilling well and probably prevents sediment from blocking the inlet to the stilling well. The channel at the stilling well location and downstream has a wide horizontal bed of natural materials, and vertical concrete side walls. Some sediment deposition on the inside of a bend approximately 300 m downstream was observed during site visits, and this is likely to be the main influence on the water level at the gauge, particularly during low flows. The sediment may be scoured during high flows, although deposits may be formed again as the flood subsides. The bridge on Jalan Ipoh is likely to have a strong influence on water levels at the gauge during high flows.

3116430 Klang River At Sulaiman Bridge

Station	3116430 Klang River at Sulaiman Bridge
Location	Approximately one kilometre downstream of Masjid Jamek, which is located at the confluence of the Gombak and Klang Rivers.
Catchment Area	468 km2
Equipment	Gas purge (bubble) pressure sensor with chart recorder, analog-digital converter providing input to telemetry system.
	Cableway with double-drum winch for stream gauging. The gauging weight is 66 kg. The current meter is an Ott with low velocity and high velocity propellers. A sediment sampler is also located at the station.
Features	The gauge is located approximately 30 m upstream of Sulaiman Bridge. The channel form at the gauge consists of a low flow channel with vertical concrete walls, berms stabilised by large rocks (300-600 mm) and vertical concrete walls above the berm.

8.3.3 Flood Related Data

DID is the custodian of a state-wide collation of flood related data that has been captured and/or developed through a number of flood risk assessment investigations and studies. The information is stored as a series of GIS layers and flood reports.

Some related information includes:

    i.   Flood-related studies
    ii.   Flood level and flood extent measurement at flood monitoring station
    iii. New or revised survey of flood extents and levels
    iv. New flood information becoming available.

The DID dataset is the reference point for all users looking for the most up-to-date Statewide flood information version.

8.4   Urban flood model

8.4.1 Model set-up

The model of the study area contains the river network and urban floodplains for which the 1D/2D commercial software packages MIKE11/MIKE21 (i.e., MIKEFLOOD) were utilised. The MIKE 11 model of the river was developed and calibrated previously by DHI Water and Environment in 2004 (See details in Appendices B and C). The drainage channel network and the river are modelled with 1D model (MIKE 11). The calculated subcatchment discharges are introduced as lateral or concentrated inflows into the branches of the 1D model network. The hydrographs generated for each subcatchment are calculated using the NAM and nonlinear reservoir method. Note that the NAM model was used only in the model developed by DHI and has not been used in the present model. Also no sewer model was considered in this current research.

The floodplain flows are modelled with the 2D model (MIKE 21). The DTM along the channel network, providing the interface between the coupled 1D-2D models, is set to the bank-full level of the 1D model. Upstream and downstream boundary conditions of the river model are derived from the results of the previous Klang River model and introduced as inflows (upstream) and water level (downstream) for the version of the model used in the present work. The flood extent resulting from the developed models is due to the combined effects of river-related overbank discharges as well as discharges from the inland drainage system.

MIKE11 Input

Network

The network adopted for the current research is shown in Figure 8.4. The chainage for Sg. Klang starts at 0 upstream and at 2952 downstream, while the chainage for Sg. Gombak starts at 0 upstream and at 1488 for the confluence with Sg. Klang.

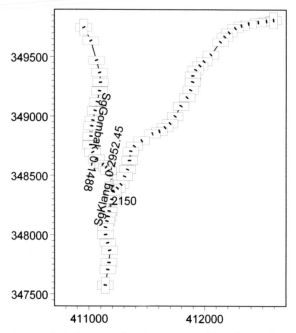

Figure 8.4: Network for Sg. Klang and Sg. Gombak

Cross-Section

The general cross-sections for both Sg. Klang and Sg. Gombak are shown in Figure 8.5 and Figure 8.6 respectively.

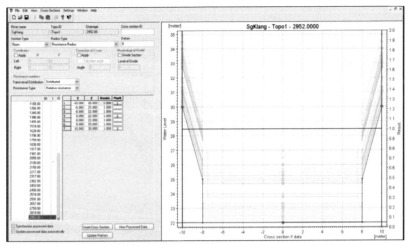

Figure 8.5: General cross-section for Sg. Klang

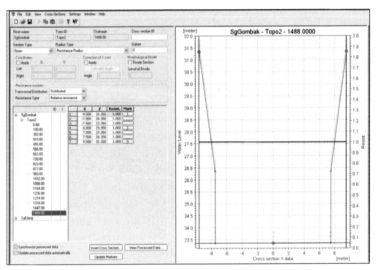

Figure 8.6: General cross-section for Sg. Gombak

Boundary Conditions

The boundary conditions are the drivers for the model. In this case the main input to the 2D model is the link to the hydrological model results. The model boundaries are listed in Table 8.2.

Table 8.2 : Boundary Conditions for 2D Model

Branch River name	Chainage (m)	Boundary Conditions Type	Assign condition
SgKlang	0.00	Discharge	0 m3/s
SgKlang	2952	Q-h	See figure 8.7

Branch River name	Chainage (m)	Boundary Conditions Type	Assign condition
SgGombak	0.00	Discharge	0 m3/s

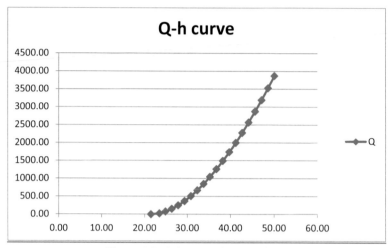

Figure 8.7: Q-h curve for Sg. Klang

Rainfall Events

The models were simulated using two historical rainfall events (29th October 2001 and 10th June 2003) that caused severe floods within the study area. The model results were analysed in terms of flood depth and extent of flooded area, and compared with recorded data from several locations. The data was obtained by the Drainage and Irrigation Department of Malaysia (DID) which made the data available for this study. The rainfall data used in model simulations was gathered from two rainfall stations: JPS Wilayah and Leboh Pasar. For the rainfall event that occurred on 29th October 2001, the JPS Wilayah station recorded 89mm in three hours while 125mm in three hours was recorded at the Leboh Pasar station. For the event that occurred on 10th June 2003 the recorded rainfall at Station JPS Wilayah was 86mm in four hours while the Leboh Pasar station recorded 125mm in four hours. The data from both stations was used in the model simulations to reflect the spatial variability of the rainfall.

MIKE 21 Flow Model

Bathymetry

The bathymetry of the urban surface for the 2D model is defined by the DTM. In this research the adopted grid is 1m x 1m. Figure 8.8 shows a screen shot of the DTM with the 1m x 1m grid size being input to the model.

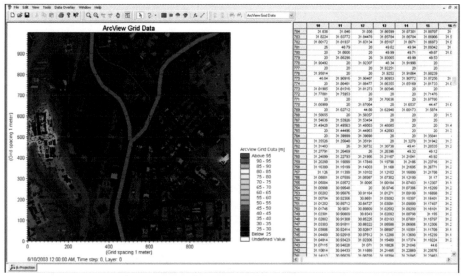

Figure 8.8 : Urban DTM with 1m x 1m grid

Simulation period

In this research, two simulation periods have been used based on the two rainfall events. The first simulation period is from 1pm to 6pm on 29 October 2001 while the other is from 3pm to 8pm on 10 June 2003. The time step used for the models is 0.1sec. Figure 8.9 shows an example of simulation period for 10 June 2003 flood event.

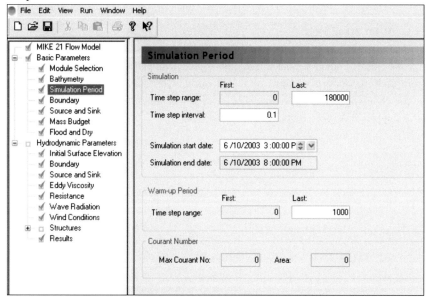

Figure 8.9 : example of simulation period for 10 June 2003 flood event

117

<u>Resistance</u>

In this research, the surface roughness coefficient (Manning's value) adopted for the MIKE21 model is 30 (n=0.033) for the entire urban surface except for the cells that represent passage buildings and close-to-earth vegetation. The Manning value of 20 (n=0.05) is used for a passage building to emulate disturbances due to local obstacles (e.g., shops, bike racks, etc.) and the Manning's value of 40 (n=0.025) is used for close-to-earth vegetation. Figure 8.10 shows an example of the distribution of the roughness coefficient for part of study area where the three Manning's values are included.

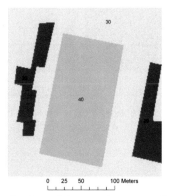

Figure 8.10: Example of the distribution of the roughness coefficients for part of the study area where all three Manning's value are included.

MIKEFLOOD 1D-2D models

The coupled 1D-2D models simulate the flow in the Klang River and its tributaries and the overtopping of flow along the streets of Kuala Lumpur. Figure 8.11 indicates how the cross-section in the 1D model is linked to the DTM along the channel network, providing the interface between the coupled 1D-2D models.

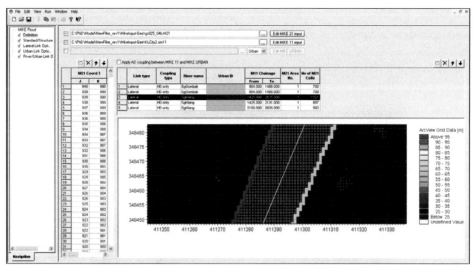

Figure 8.11 The link between 1D cross-section and the DTM along the channel network

Interpolation of river banks

The reason for the interpolation of the river banks comes from the relation between the 1D and 2D elements that is adopted in the 1D/2D model for this study. Figure 8.12 shows the 1D and 2D elements in an area drained by an open channel. The channel is represented by the cross sections typical of a 1D unsteady flow model, while the flood plain is represented as a 2D grid which is provided by the DTM. The elevations of the 1D nodes represent the top of the banks for the 1D channel. The surface area of the 1D model is cut out of the 2D grid so that a double accounting of the flow volume is not permitted. The water surface in 1D is allowed to rise vertically in the channel, and this same water level is tied to the water level in adjacent water cells as defined in the 1D/2D model. While the river depth and width has been modelled by the 1D unsteady flow model the representation of the river is not needed in the DTM.

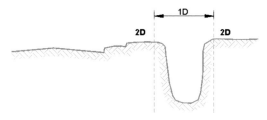

Figure 8.12: Section of channel showing 1D and 2D elements in a 1D/2D linked model

1D/2D coupling model

The DTM required for the 2D hydraulics model in this study is a very detailed topography of catchment surface. The adopted grid is 1m x 1m resolution with vertical accuracy of 0.15m. Optimization of the performance of the 2D flood models has been done by the automated incorporation and removal of flood diverting landscape elements such as buildings, elevated roads and bridges. In the case of piles with a width of more than 1m, it is not a problem for them to be incorporated in the DTM. The 2D hydraulics models were constructed from the different DTMs and coupled with the 1D model.

Site Investigation

In order to accommodate the Phase I of MPMA development, a field survey was carried out in order to classify the buildings into those that have basements, passage buildings and those that can be regarded as solid objects. Out of 191 surveyed buildings 51% were found to be the buildings with basements, 43% passage buildings and 6% solid object buildings (see Figure 8.13).

Figure 8.13: Example of a building with basement (left), a solid object building (top right corner) and a passage building (bottom right corner).

The analysis of survey data indicates:

   i.   92% of buildings with a height in the excess of 20 meters have basements
   ii.   85% of buildings with a height between 5 to 20 meters have a significant open space on the ground floor
   iii. 87% of solid objects which are less than 5 meters high have neither basements nor open space on the ground floor

For buildings with a basement, the average basement depth was found to be the order of 5 meters. The values gathered from this survey are used to define the characteristics of study area.
A field survey was also carried out in the study area in order to gather information about the piles beneath elevated roads and to use such information for incorporation in a DTM (Phase II). Survey data concerning typical pile geometry in the study area can be summarised by the following data:

i.   Height, h = 6 m
ii.   Length, l = 2 m
iii.   Width, w = 6 m
iv.   Distance between piles, lb = 32m

8.4.2 2D Hydrodynamic Model calibration

The calibration process of the 2D hydrodynamic model is a trial and error iterative process, during which a set of modelling parameters are adjusted until the simulated water levels and discharges in the river system are in good agreement with the observed data at the selected gauging stations for the rivers considered in the model. The main parameter for calibration is the Manning number.   In a one dimensional mathematical model such as MIKE 11 HD, this parameter lumps together in a single value a number of hydraulic phenomena as contributing to the roughness of the river bed, such as bed forms, the sinuosity of the river, natural obstacles, etc. However, in complex river systems such as the Sg. Klang which includes many man-made river flow obstructions such as bridges, regulation structures, dams, etc, local head losses and the operation of these structures are also part of the calibration procedure because the flow conditions will not only depend on the general friction losses and cross-section shape but also on these river flow obstructions. The calibration procedure is therefore a gradual learning process on how the river system behaves and fully depends on the quantity and quality of the available data.

Selection of calibration events
The events listed in Table 8.3 were selected because they occurred recently. The main concern in selecting these events is that the Klang river has undergone many changes during the last decade, particularly during the last few years.

Table 8.3: Selected Periods for calibration of the hydrodynamic model

Event No.	From	To
1	03/SEP/1998	07/SEP/1998
2	30/APR/2000	02/MAY2000
3	19/DEC/2000	21/DEC/2000
4	28/OCT/2001	30/OCT/2001
5	09/JUN/2003	10/JUN/2003

During the calibration, the simulation results are compared to the time series records from measuring stations within the model area. The following recording stations (see Table 8.4) were selected for the calibration of the model. They have been key to the calibration of the hydrodynamic model.

Table 8.4: Recording stations selected for the calibration of the model

Station No.	River Name	Model Chainage (m)	Remarks
3116430	Klang	17385	Upstream of Sulaiman bridge

Figures 8.14 to 8.23 illustrate the model simulation results at the listed gauging locations. The black continuous line indicates the model simulation results and the red continuous line with symbols represents the observations at the selected station.

Event No.1: 03-07 September, 1998

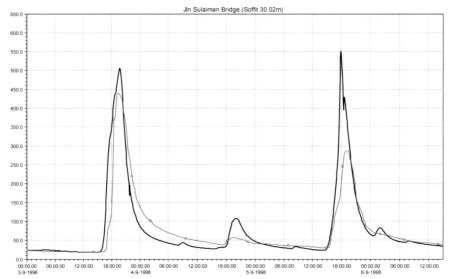

Figure 8.14:   Event No.1: Discharge (station 3116330)

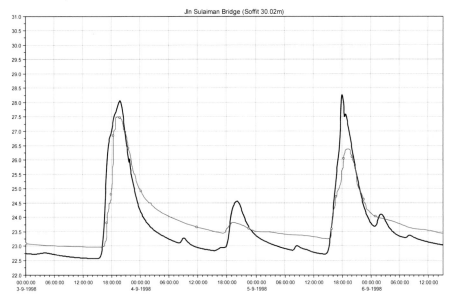

Figure 8.15   Event No.1: Water level (station 3116330)

Event No.2: 30 April – 02 May, 2000

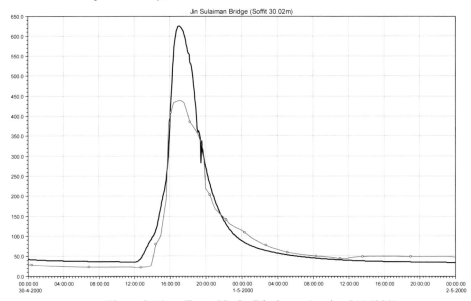

Figure 8.16:   Event No.2 : Discharge (station 3116330)

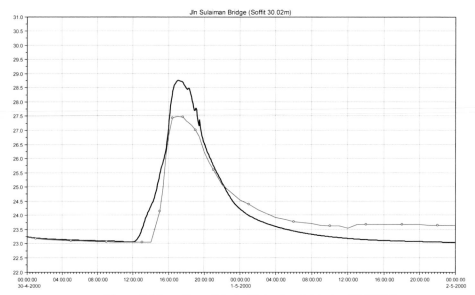

Figure 8.17: Event No.2: Water level (station 3116330)

Event No.3: 19 – 21 December, 2000

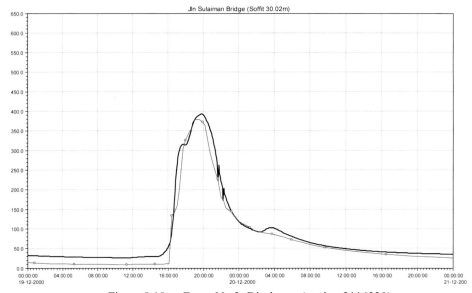

Figure 8.18 : Event No.3: Discharge (station 3116330)

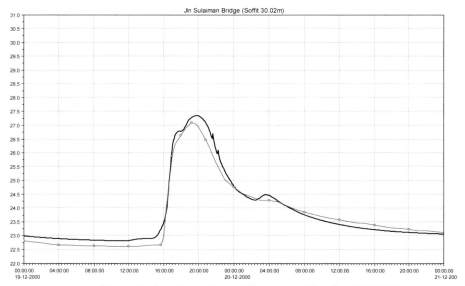

Figure 8.19 :   Event No.3: Water Level (station 3116430)

Event No.4: 27 October, 2000 – 29 October, 2001

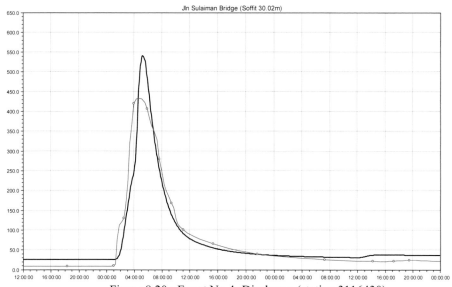

Figure 8.20 : Event No.4: Discharge (station 3116430)

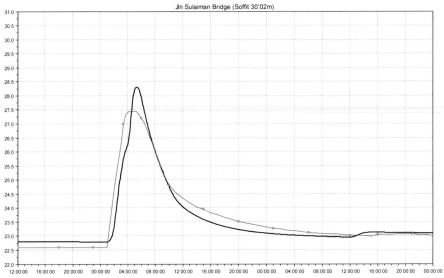

Figure 8.21 : Event No.4: Water Level (station 3116430)

Event No.5: 8 June, 2003 – 10 June, 2003

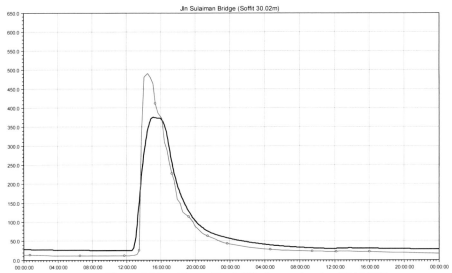

Figure 8.22:   Event No.5: Discharge (station 3116430)

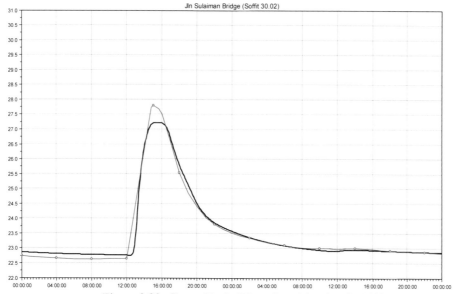

Figure 8.23 : Event No.5 : Water Level (station 3116430)

In general, the current 2D hydrodynamic model simulates the hydraulics of the system consistently. However, the hydrodynamic model should be reviewed with new update information consistent with the prototype river system when it is available as this data will clearly improve the performance of the model before it it is applied to the design and optimisation of any flood mitigation scheme or flood forecasting system.

8.5   Comparison between Progressive Morphological Algorithm (PMA) and Modified Progressive Morphological Algorithm (MPMA)

For the purpose of comparing the existing Progressive Morphological Algorithm (PMA) with the proposed filtering algorithms (MPMA) two 2D models were generated from the 1-meter grid DTMs. In context of handling the buildings, PMA is poor in capturing the buildings accurately. This is because PMA tend to remove all objects and try to preserve the bare earth as much as it can. When compared with the DTM from MPMA (shown in Figure 8.24) it is clearly seen that PMA did not include the buildings in the DTM. Almost all buildings are removed along with the vegetation. MPMA incorporates buildings together with the basement properties for those buildings with a basement and incorporate the different roughness coefficient (Manning's value) for passage buildings.

127

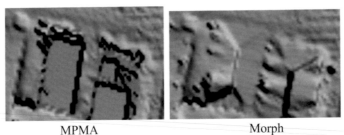

MPMA   Morph

Figure 8.24: Urban DTM from MPMA and PMA respectively

From the DTM, it is also seen that there are likely to be some differences in floodwater flow with both DTMs when flooding occurs. Referring to Figure 8.8 which shows the DTM for PMA and MPMA, when floodwater flow confronts buildings with basements in the area for PMA's DTM, most of the flow is diverted or passes through the buildings resulting from the building 'ruin' which has not been removed completely in filtering process. In MPMA, the floodwater flows into the basement first before it floods another area. This difference in the flow behaviour has a considerable impact on the flood depth and flood extent results.

In relation to the treatment of elevated roads/train lines, the comparison of the DTMs produced by PMA and MPMA2 algorithms was undertaken at two typical locations (Figure 8.25).

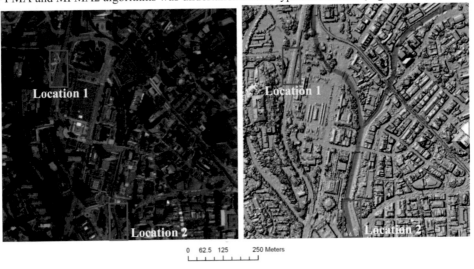

0   62.5   125   250 Meters

Figure 8.25: Sample location within the study area

Location 1 shown in Figure 8.25 is situated in the North-West of the study area and has many elevated roads. Location 2 is situated at the downstream end of the study area and contains a typical bridge structure. A comparison of the DTMs produced by the two algorithms is given in Figure 8.26. A qualitative assessment was undertaken to evaluate PMA and MPMA2. In this

128

assessment, criteria were used that focus on the removal of elevated roads and bridges. The assessment was done by visually analysing the performance and giving a mark with a weighted value. This weighted value is based on the filter performance in removing the features. If more than 75% of the features are removed the filter is given 1 mark; if 50% to 75% of the features are removed the filter is given 2 marks; if 25% to 50% of the features are removed the filter is given 3 marks and if less than 25% of the features are removed the filter is given 4 marks. The least total mark gives an indication which filter performs best with respect to the selected criteria. The DTMs were compared in terms of their efficiency in removing elevated roads and bridges in the two locations, see Figure 8.25. Further to the improvements described above, the DTM generated by MPMA algorithm proved to give better results than its predecessor PMA. The DTM generated by MPMA appeared as almost flat in all two locations and it managed to incorporate piles at location 1. PMA could remove only 40% to 70% of the features, showing that the two filters act differently and have different capabilities. From the overall analysis of results given in Table 8.5 it can be noted that the MPMA produced better results than PMA. This is also confirmed by observing Figure 8.26, which illustrates the quality of DTMs obtained by two algorithms at the two locations (one at the location with elevated road and the other at the location with the bridge structure).

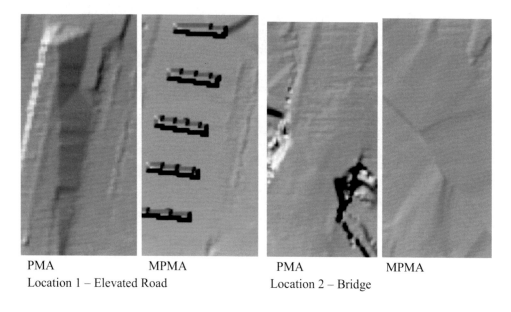

| PMA | MPMA | PMA | MPMA |

Location 1 – Elevated Road   Location 2 – Bridge

0   5   10   20 Meters

Figure 8.26: Comparison of DTM details produced by two algorithms at two locations.

Table 8.5: Comparison of performance by two algorithms.

Indicator	PMA	MPMA
Bridge Removal	2	1
Elevated Road Removal	3	1
River Alignment	2	1
Building handling	1	1
Total Weighted Value	**8**	**4**

d Value: 1=Excellent, 2=Fair, 3=Acceptable, 4=Poor,

In relation to the recovery of curbs, a comparison of the DTMs produced by PMA and MPMA algorithms was undertaken at three typical locations (Figure 8.27). Location 1 to Location 3 in Figure 8.27 is situated in the South-East of the study area that has many road networks. A comparison of the DTMs produced by the two algorithms is given in Figure 8.14 and discussed below.

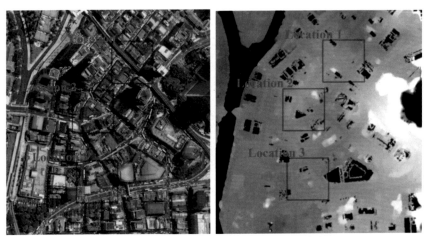

Figure 8.27: Sample location within the study area

A qualitative assessment was undertaken to evaluate the DTM developed by PMA and MPMA. In this assessment, the focus is on the recovering the curbs. The assessment was done by visually analyzing the DTM and giving a mark with a weighted value. This weighted value is based on the performance in recovering the curbs. If more than 75% of the curbs are captured/recovered the filter is given 1 mark; if 50% to 75% of the curbs are captured/recovered the filter is given 2 marks; if 25% to 50% of the curbs are captured the filter is given 3 marks and if less than 25% of the curbs are captured/ recovered the filter is given 4 marks. The least total mark gives an indication which filter performs best with respect to the selected criteria.

Table 8.6: Comparison of performance in curbs capturing and recovering by two algorithms.

Indicators	PMA	MPMA
Curbs captured/recovered	3	1
Discontinuity problem	3	1
Total Weighted Value	6	2

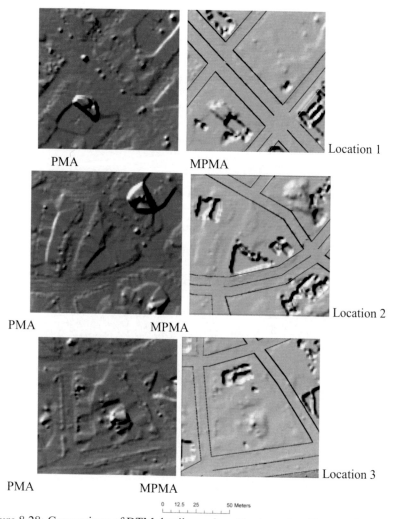

Location 1

PMA   MPMA

Location 2

PMA   MPMA

Location 3

PMA   MPMA

0   12.5   25   50 Meters

Figure 8.28: Comparison of DTM details produced by two algorithms at three locations.

The DTMs were compared in terms of their efficiency in removing recapturing and recovering the curbs in three locations, Figure 8.27. From the overall analysis of results given in Table 8.6 it can be seen that MPMA produced better results than PMA. This is also confirmed by Figure 8.28, which illustrates the quality of the DTMs obtained by the two algorithms at the three locations. Further to the improvements described above, the DTM generated by MPMA algorithm proved to give better results than its predecessor PMA. The DTM generated by MPMA appeared to recover almost all the curbs that have been left out due to an insufficient grid resolution in PMA (1m) at those locations. MPMA managed to incorporate curbs in all three locations using the sub-grid method which in this case was upgraded to 0.2m (only at locations where curbs have been recovered). PMA could capture less than 20% of the curbs enclosed due to curbs discontinuity problem which can be clearly seen in all locations.

8.5.1 Quality Analysis

In order to determine the potential influence of the filtering algorithms on the resulting DTM, Type I (bare earth misclassified as objects) and Type II (objects misclassified as bare earth) errors in the LiDAR information extraction is calculated. The calculation error is done by calculating the percentage of Type I and Type II errors based on the predominant features in the data set.

Five test sites in urban environment within the case study area (See Figure 8.29) were chosen because they contained a variety of characteristics that were expected to be difficult for automatic filtering. The datasets included landscapes with densely packed buildings, elevated roads and underpasses, river and bridges, ramp and curbs and large area of bushes and grass. The urban sites were recorded with a point spacing of 1m.

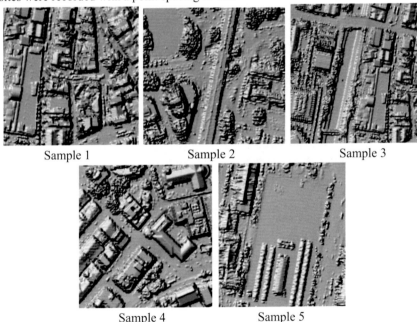

Sample 1   Sample 2   Sample 3

Sample 4   Sample 5

Figure 8.29: Test sites

All points in the datasets were labeled bare earth or object. The definition of what should be considered bare earth is, however, a subjective one. For the purpose of this test, bare earth was defined as the topsoil or any thin layering (asphalt, pavement, etc.) covering it. According to this definition, buildings, bridges, gangways, etc., were treated as objects. Additionally, the bare earth was treated as being a piecewise continuous surface. Therefore, courtyards were also accepted as being part of the bare earth if they were near the surface interpolated between the points on surrounding streets. From the two data sets (PMA and MPMA) five samples were abstracted. These five samples were representative of different selected important elements in urban environments. They were focused in respect to the expected difficulties in handling those elements as identified in Chapter 7.

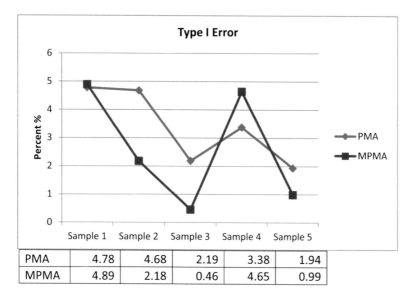

	Sample 1	Sample 2	Sample 3	Sample 4	Sample 5
PMA	4.78	4.68	2.19	3.38	1.94
MPMA	4.89	2.18	0.46	4.65	0.99

Figure 8.30: Type I errors over the 5 samples

The type I errors calculated are shown in Figure 8.30. The type I error presents the number of misclassified bare earth points in a sample as a percentage of all bare earth points in the sample. The developed algorithm does not exhibit large error variations in type I errors (0-5%) between the sample sites. This is encouraging because it indicates that the developed algorithm is more robust to different landscape types and hence is more reliable.

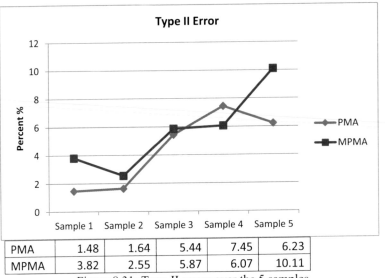

	Sample 1	Sample 2	Sample 3	Sample 4	Sample 5
PMA	1.48	1.64	5.44	7.45	6.23
MPMA	3.82	2.55	5.87	6.07	10.11

Figure 8.31: Type II errors over the 5 samples

The type II errors calculated are shown in Figure 8.31. The type II error presents the number of misclassified object points in a sample as a percentage of all object points in the sample. The type II errors obtained were relatively small, except for a few sites where the solid buildings and curbs are preserved and rebuilt in the developed algorithm which led to higher errors. Usually, there are more bare earth points then object points which make the impact of type II errors on the total error is small.

The results for each are discussed in more detail.

Sample 1: The developed algorithm performed very well on this site. The flat bare earth and well elevated buildings make this a relatively simple landscape to filter. Type I errors arise from the points that represent the solid building which on purposely preserved in the developed algorithm. Because of the preservation, these points are classified as bare earth points. The increased point density in the overlap regions leads to under estimating in classifying the large object. Hence type II, errors typically appear in the overlap regions.

Sample 2: The special characteristic of this sample are the elevated road and under passes. There are no problems in detecting the large elevated road. After the detection the elevated road detection algorithm using the combination with the LiDAR intensity value was run and the rest of the elevated roads were detected. The elevated road was slightly over detected, hence leading to type I errors. Other type I errors were caused by the micro object detection which mis-classifies low pavement edges. Type II errors were in this sample is caused by low vegetation.

Sample 3: The special characteristic of this site are the river and the bridges. In the filtering, the bridge was successfully detected and later removed and led to the very low Type I errors. Type II errors are mainly result from the interpolation between river banks which make the rivers and also bridges is classified as bare earth. Despite this the error is kept small and importantly for an urban scene discontinuities are preserved.

Sample 4: This small sample show the effect of curbs in the landscape. Curbs are typically treated as part of the bare earth but in the developed algorithm, it is treated as objects. The width of the curbs is less then 1m and hence it was barely detected and when detected it was overshadowed with the discontinuity problem. Most of the type I errors result from misclassification of pavement. This is also the cause of the errors at the base of the curbs. Type II errors are quite high and most of them occur in the areas where there curbs are rebuilt using the sub-grid method.

Sample 5: This sample contains a data gap, road embankments, ditches and area with bushes and grass. The challenge in this data set is to filter a landscape that contains relatively few objects. In this the algorithm does very well with a type I error of only 1%. The type II error is caused by low vegetation points on the road embankments.

8.6 Evaluation of MPMA using the Urban Flood Model

The evaluation of the proposed filtering algorithm MPMA is done by applying the DTM generated using the filtering to particular flood models. The results from the flood models are compared for differences in four variables, namely flood depth, flood extent, flood velocity and distribution of surface roughness. In this research, the development of the proposed filtering algorithm, MPMA is done step by step, which divides the whole process into three phases. Each phase focuses on one or two related selected urban elements that are essential in the prediction of urban flooding using an urban flood model; buildings, elevated road and rail line, curbs and closed-to-earth vegetation. Each urban element is handled by some specific approach which in the final stage is combined with the other approaches dealing with other elements into one modified filtering algorithm that hopefully can best suit the requirements of an urban flood model.

8.6.1 Flood Depth

The evaluation of flood depths is done by comparing the results from urban flood models with the measurement flood depth value recorded by Department of Irrigation and Drainage Malaysia. As the development is done in phases, the evaluation is also conducted in phases in which every element is discussed separately.

Phase I: Detection and modification of buildings (MPMA1)

The evaluation in this Phase I is done by comparing the results from urban flood models derived from developed algorithm, MPMA with 4 others existing filtering algorithm in term of flood depth. As shown in Table 3, for the event that occurred on 29th October 2001, it can be observed that there are significant differences between the computed and recorded data. For the Morph, Morph2D, Polynomial and Adaptive TIN filtering algorithms, the corresponding 2D models give a reasonable prediction of flood depths. This can be attributed to the small difference in features filtered by the different algorithms and also the lack of capability in distinguishing buildings with a basement. The effect of floodwater entering and ponding in the basement of a building first before continuing to flow into the surrounding area is obviously a critical aspect, and the models which do not have such a representation are unlikely to produce appropriate results, especially for buildings with sizeable basements that are not flood-proofed. The difference in the developed algorithm is shown in the DTM where the buildings are categorized in three groups and treated differently depending on their contribution to diverting flood flows. It seems that because of this, and for other reasons such as the difference in roughness coefficient for a passage building compared to the overall area, that the results from MPMA appear to present a more reliable flow movement and lead to a closer match with the measurements. The model results for flood depths are shown in Table 8.7.

Table 8.7: Summary of modelled and measured flood depths for 29th October 2001 rainfall event.

Location / Flood depth (m)	Dataran Merdeka (1)	Leboh Ampang (2)	Tun Perak (3)	Jalan Melaka (4)	Kg. Pantai Halt (5)
MEASURED	0.370	1.870	0.870	1.370	0.540
1D/2D model with MPMA DTM	0.410	1.656	0.745	1.238	0.949
1D/2D model with POLY DTM	0.599	1.303	0.599	1.702	1.540
1D/2D model with MORPH DTM	0.532	1.240	0.361	1.617	1.380
1D/2D model with MORPH2D DTM	1.207	0.825	0.353	2.134	1.350
1D/2D model with ATIN DTM	1.165	1.033	0.403	1.135	1.186

Table 8.8 shows the summary of modelled and measured flood depths for the flood event that occurred on 10th June 2003. From this table it can be observed that the model which uses the DTM built from the MPMA algorithm once again gives the closest match to the measurements. This can be explained by the fact that in MPMA DTM the buildings are more adequately represented.

Table 8.8: Summary of modelled and measured flood depths for 10th June 2003 rainfall event.

Location / Flood depth (m)	Dataran Merdeka (1)	% Error	Leboh Ampang (2)	% Error	Tun Perak (3)	% Error	Jalan Melaka (4)	% Error
MEASURED	0.500		1.200		1.000		1.300	
1D/2D model with MPMA DTM	0.923	84.60	1.192	0.67	0.964	3.60	1.621	24.69
1D/2D model with POLY DTM	1.250	150.00	1.219	1.58	1.067	6.70	1.822	40.15

136

1D/2D model with MORPH DTM	1.471	194.20	1.254	4.50	1.180	18.00	1.910	46.92
1D/2D model with MORPH2D DTM	1.469	193.80	1.321	10.08	1.195	19.50	1.995	53.46
1D/2D model with ATIN DTM	1.392	178.40	1.210	0.83	0.885	11.50	1.792	37.85

Location Flood depth (m)	Jalan Parlimen (5)	% Error	Jalan Raja (6)	% Error	Leboh Pasar (7)	% Error	Jalan HS Lee (8)	% Error
MEASURED	0.500		1.000		0.650		1.000	
1D/2D model with MPMA DTM	0.855	71.00	0.877	12.30	0.748	15.08	0.996	0.40
1D/2D model with POLY DTM	1.240	148.00	1.588	58.80	1.071	64.77	1.005	0.50
1D/2D model with MORPH DTM	1.395	179.00	1.787	78.70	1.319	102.92	0.911	8.90
1D/2D model with MORPH2D DTM	1.402	180.40	1.859	85.90	1.363	109.69	1.061	6.10
1D/2D model with ATIN DTM	1.153	130.60	1.666	66.60	1.260	93.85	0.885	11.50

Phase II: Detection and modification of elevated road and bridges (MPMA2)

The evaluation of Phase II is done similarly to that for Phase I, but it is only done for one flood event, 10th June 2003. As shown in Table 8.8, it can be observed that there are significant differences between the computed and recorded data: in some locations (for example at Jalan Melaka) the difference between the measured and computed flood depths for some models is almost 0.8m. This can be attributed to the difference in features filtered by the different algorithms, and also the inability of some methods to completely remove elevated roads and to handle the river banks. The average error is calculated by excluding the results from Dataran Merdeka (1) and Jalan Parlimen (5). This is because the error is found to be quite high in these locations. One reason for this is that both areas are situated in the North-West of the study area where many obstacles, such as elevated roads and buildings, are located near the river. This situation creates a high degree of discontinuity in the DTM due to the inability of the filtering algorithms to remove unwanted objects and therefore it creates more local depression and storage areas. Furthermore, these two locations are located too far from the rainfall station, which contributes to an overestimation of the models results due to the high spatial variability. The results show that the model using the MPMA2 DTM produces results with the least error (less than 4%) (see Table 8.9) while the results from other models excluding MPMA1 are in the range 28% - 48%. Overall, it can be noted that the results from MPMA1 and MPMA2 are very close to the measurements. The differences between MPMA1 and MPMA2 with the measured value [ABS(MPMA1-Measured)/Measured, ABS(MPMA2-Measured) /Measured] is 3.80% and 3.59% respectively showing that further improvements have been achieved by MPMA2, compared to MPMA1. In this evaluation it is seen that the close to complete removal of the elevated highway resulting from the combination of height and slope difference with LiDAR intensity value adopted in MPMA in the case study area can really contribute to a more realistic flow movement in the flood event. This is one of the reasons why the result from MPMA's flood model is very close to the measurement.

Table 8.9: Summary of modelled and measured flood depths, for the 10th June 2003 rainfall event.

Location	Dataran Merdeka (1)	% Error	Leboh Ampang (2)	% Error	Tun Perak (3)	% Error	Jalan Melaka (4)	% Error
Flood depth (m)								
MEASURED	0.500		1.200		1.000		1.300	
1D/2D model with MPMA2 DTM	0.871	74.20	1.198	0.17	0.984	1.60	1.548	19.08
1D/2D model with MPMA1 DTM	0.923	84.60	1.192	0.67	0.964	3.60	1.621	24.69
1D/2D model with POLY DTM	1.250	150.00	1.219	1.58	1.067	6.70	1.822	40.15
1D/2D model with MORPH DTM	1.471	194.20	1.254	4.50	1.180	18.00	1.910	46.92
1D/2D model with MORPH2D DTM	1.469	193.80	1.321	10.08	1.195	19.50	1.995	53.46
1D/2D model with ATIN DTM	1.392	178.40	1.210	0.83	0.885	11.50	1.792	37.85

Location	Jalan Parlimen (5)	% Error	Jalan Raja (6)	% Error	Leboh Pasar (7)	% Error	Jalan HS Lee (8)	% Error
Flood depth (m)								
MEASURED	0.500		1.000		0.650		1.000	
1D/2D model with MPMA2 DTM	0.793	58.60	0.933	6.70	0.734	12.92	0.980	2.00
1D/2D model with MPMA1 DTM	0.855	71.00	0.877	12.30	0.748	15.08	0.996	0.40
1D/2D model with POLY DTM	1.240	148.00	1.588	58.80	1.071	64.77	1.005	0.50
1D/2D model with MORPH DTM	1.395	179.00	1.787	78.70	1.319	102.92	0.911	8.90
1D/2D model with MORPH2D DTM	1.402	180.40	1.859	85.90	1.363	109.69	1.061	6.10
1D/2D model with ATIN DTM	1.153	130.60	1.666	66.60	1.260	93.85	0.885	11.50

Phase III: Detection and modification of curbs and close-to-earth vegetation (MPMA3)
In this Phase III, the evaluation is done between the MPMA2 and the complete version, MPMA3. This is because MPMA2 has the same characteristics in terms of the detection of curbs as other existing filtering algorithm (which are introduced in this research).The differences between the measurements and results from the 2D model built with MPMA3 were found to be between 0.4% and 4.4%. In terms of the differences between the measurements and results from the 2D model built with the MPMA2 algorithm, this was found to be in the range 0.2% to 6.7%. In Jalan Raja (1) and Tun Perak (3), the differences between the measurements and results from

the 2D model built with MPMA3 seem to be smaller and closer to the measurement. In Leboh Ampang (2), the results show an increase in flood depth by 0.011m of the corresponding depths from the 2D model built with MPMA3, while in Jalan HS Lee (4) the result is the same for both models. The increase in flood depths at Leboh Ampang (2) can be attributed to the effects of the curb recovery along the streets, which encourages the flood water to gather more at the center of those streets. The effect of close-to-earth vegetation in this area is found to be less significant because the covered area is relatively small compared to the urban surface (streets, pavements). Table 8.10 indicates the difference between the measured and modelled data at these locations.

Table 8.10: Summary of modelled and measured flood depths, for the 10th June 2003 rainfall event.

Location	Jalan Raja (1)	% Error	Leboh Ampang (2)	% Error	Tun Perak (3)	% Error	Jalan HS Lee (4)	% Error
Flood depth (m)								
MEASURED	1		1.2		1		1	
1D/2D model with MPMA3 DTM	0.956	4.4	1.209	0.7	0.996	0.4	0.995	0.5
1D/2D model with MPMA2 DTM	0.933	6.7	1.198	0.2	0.984	1.6	0.995	0.5
1D/2D model with MPMA1 DTM	0.877	12.30	1.192	0.67	0.964	3.60	0.996	0.40
1D/2D model with POLY DTM	1.588	58.80	1.219	1.58	1.067	6.70	1.005	0.50
1D/2D model with MORPH DTM	1.787	78.70	1.254	4.50	1.180	18.00	0.911	8.90
1D/2D model with MORPH2D DTM	1.859	85.90	1.321	10.08	1.195	19.50	1.061	6.10

8.6.2 Flood Extent

In term of flood extent, the evaluation is done by comparing the model results with the measurement of flood extent recorded by the Department of Irrigation and Drainage Malaysia (DID). The DID used several approaches to record the flood extents. One of these methods consists of a site investigation during the event, and the other method is by using remote sensing where the flood extent is digitized as seen on the satellite images. The second method is usually difficult to do because on the day of the event, there is normally a thick coverage of clouds throughout the area, which obscures the flood extent. One solution to this problem is to use radar data (e.g. Radar-sat) because unlike the common satellite images, the sensor used in a radar system can penetrate the cloud coverage.

Phase I: Detection and modification of buildings
Similar to the evaluation of flood depth in Phase I, this evaluation is done by comparing the results from urban flood models derived from the developed algorithm, MPMA, with 4 other existing filtering algorithms but this time in terms of extent of flooding for the event on 10th June

2003. For the analysis of flood extent in addition to the measurements taken at several locations by DID, local observations were also sourced from newspapers such as Utusan Malaysia, The Sun, New Strait Times, The Star, and Harian Metro and Berita Harian. These data were introduced into the GIS layers and then overlaid with the model results for comparison purposes. Figure 8.32 illustrates flood extent produced by five 1D/2D models against local observations. From Figure 8.32 it can be observed that the flood extent obtained from all the models is in fairly good agreement with the measurements, but the flood extent from Morph, ATIN, Morph2D and Poly appears to be larger than what was recorded. Again, the model results based on the MPMA DTM were found to be closer to local observations than other model results.

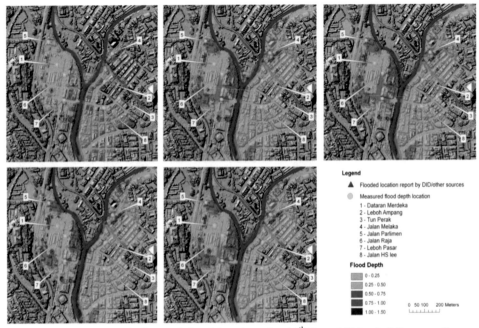

Figure 8.32: Modelled and observed flood locations for 10th June 2003 rainfall event. Top row illustrates predictions by models with DTMs generated from MPMA, Morph and Morph2D algorithms. Bottom row illustrates predictions by models with DTMs generated from Poly and ATIN algorithms. Observed locations are represented with triangles and circles.

Phase II: Detection and modification of elevated road and bridges
In Phase II, the evaluation is done by comparing the results in term of flood extent from urban flood models derived from MPMA1 with MPMA2.

Figure 8.33: Location for flood extent comparison between MPMA1 and MPMA2

Continuing the analysis of flood extents, it can be seen that there are differences between the models using MPMA1 and MPMA2, especially in places where elevated roads and bridges have been removed. Figure 8.33 shows three different locations in the study area, where a difference in flood extent is detected between MPMA1 and MPMA2. Each location covers an area of 6.8ha (0.068km^2). At Location A (see Figure 8.34), the flood extent appears to be more widespread in the model that uses a DTM based on the MPMA2 algorithm (with approximately 75% coverage of the area), compared to the results of the model that uses a DTM based on the MPMA1 algorithm (with approximately 60% coverage of the area). This is mainly due to the presence of the elevated road, which has been completely removed from the DTM in MPMA2, thus allowing the water to flow through. A similar result was detected at Location C (see Figure 8.34), where the elevated train line was removed. At Location B (see Figure 8.34), because of its location downstream, the removal of bridges and even an elevated road or train line from the upstream area influences the flood extent to be wider but shallower in MPMA2 compared to MPMA1 which has a more restricted flood extent (less extent but greater depth). In an area of 6.8ha, approximately 3.1ha is covered by the flood in MPMA2 and only 2.0ha is covered in MPMA1. Overall, the flood extent results show that models are capable of producing simulations close to local records.

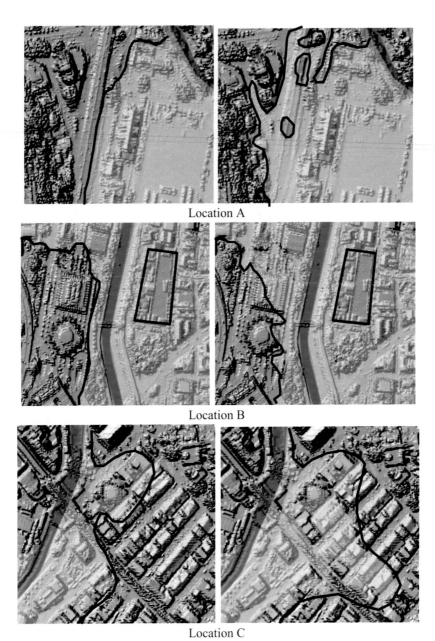

Location A

Location B

Location C

Figure 8.34: The difference between flood extents in MPMA1 (left) and MPMA2 (right). The flood extents have been superimposed on the top of the original Digital Surface Model (DSM) of study area.

Phase III: Detection and modification of curbs and close-to-earth vegetation
The evaluation in Phase III, similar to that done for the evaluation in Phase II, compares the

results of flood extent for the urban flood models derived with MPMA2 and MPMA3 for the event of 10th June 2003. In the study area ($1km^2$), the results from the 2D model built with the MPMA2 algorithm show that 30% of that area (or 300,000 m^2) is flooded. In terms of the results obtained from the 2D model built with MPMA3 algorithm, the flooded areas (315,000 m^2) are increase by approximately 5% compared to the model built with MPMA2. This result shows that the presence of rebuilt curbs in the DTM and the difference in the roughness coefficient for the close-to earth vegetation (grass) have only a minor influence on the difference in flood extent.

8.6.3 Flow Velocity

In terms of flood flow velocity, the evaluation is done by visually observing the flow direction and flow speed in the study area when the flood happens using the flood event of 10th June 2003.

Phase I: Detection and modification of buildings

In Phase I, similar to the evaluation done for flood depth and flood extent, the comparison is done with the existing filtering algorithm. This is because in the first phase, the efficiency of the developed algorithm can only be compared with the existing filtering algorithm. From the analysis of computed flood velocities it can be observed that the results from 1D/2D model, which uses the DTM from the MPMA algorithm, are closer to reality as they better represent the physics of the phenomena; see Figure 8.35. This is more or less a consequence of the ability of MPMA to recreate the basement condition in the DTM where the flood water is allowed to inundate the basement area first before it floods the surrounding area. Not only that, the assigned Manning value of 20 (n=0.05) for passage buildings produces a flow that more closely represents the real phenomena. From the overall results it can be concluded that models with different DTMs can lead to significantly different flood predictions. The difference in the results suggests the importance of a careful consideration of building objects for urban floodplain modelling.

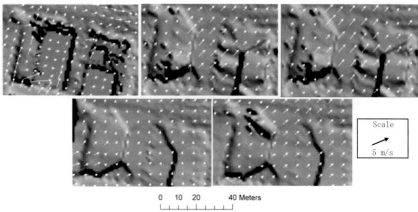

Figure 8.35: Comparison of velocity vectors. Top row shows modelled velocity vectors and DTMs generated from MPMA, Morph and Poly. Bottom row shows modelled velocity vectors and DTMs generated from Morph2D and ATIN.

Phase II: Detection and modification of elevated road and bridges

In Phase II, the comparison is done between MPMA1 and MPMA2. From the analysis of computed flood velocities, it can be observed that the results from the 1D/2D model that uses the DTM from the MPMA2 algorithm, show some improvement and come closer to reality because the DTM better represents the physics of the phenomena (see Figure 8.34). This is due to the ability of MPMA2 to completely remove elevated roads, and bridges in the DTM, where the flood water is then allowed to flow through these features. As for the removal of bridges, even though, after the removal process the water that overtops the river can flow with less obstruction, its momentum is still affected by the river channel and flow in the river. For example, at Location 1, an elevated road is present in the DTM created from the MPMA1; this is clearly shown in the cross section (see Figure 8.36 – Location 1). The flood model treats this as an impermeable mound. From the velocity vector plot, it is seen that the flood flows in this case are diverted by the obstruction, which should not happen in reality. In MPMA2, the elevated road has been removed from the DTM and the velocity vector plot shows a better representation of the physics. As for Location 3, it is seen that the bridge has been removed and the riverbank has been interpolated in the DTM from the MPMA2. From the velocity vectors, as discussed above, the flow is still affected by the river channel and flow in the river. The difference in results suggests the importance of careful processing and removal of objects from the raw LiDAR data, otherwise false obstacles in a DTM can generate misleading results.

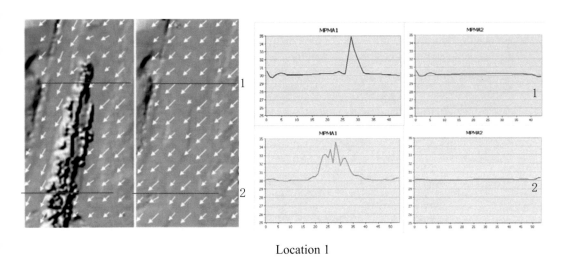

Location 1

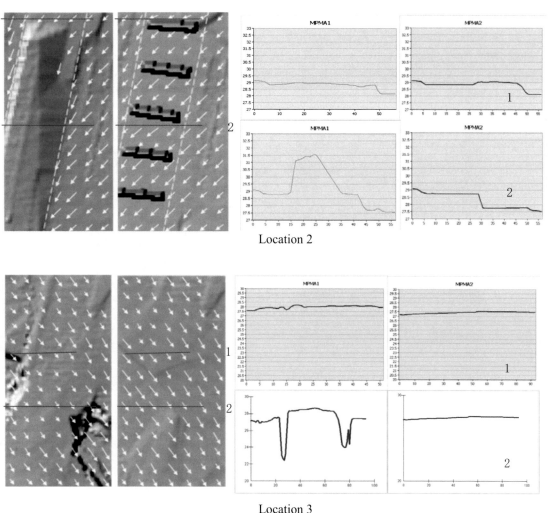

Location 2

Location 3

Figure 8.36: Comparison of velocity vectors and DTMs generated from MPMA1 (left image) and MPMA2 (right image) and cross-sections (y-axis represent the elevation in meter while x-axis represent length in meter) at three sample locations.

Phase III: Detection and modification of curbs and close-to-earth vegetation

In Phase III, the comparison is done between MPMA2 and MPMA3. From the analysis of the computed flood velocities, it can be observed that the results from the 1D/2D model that uses the DTM from the MPMA3 algorithm shows that the flood water has been controlled by the curbs to flow along the street. This result corresponds better to reality because the DTM better represents the street characteristics (Figure 8.36). In the 1D/2D model that uses the DTM from the MPMA2 algorithm the water is seen to flow arbitrarily through the streets because the curbs cannot be captured by the 1m grid resolution. This is also evident from observations of velocity vectors

145

given in Figure 8.36. As at Location A, it is seen that in the MPMA3 model, where the surface is covered by the close-to-earth vegetation, which is in this case, grass, the flood is observed to flow more slowly than over the same area in the MPMA2 model. The difference in results indicates that micro objects, especially those corresponding to the street characteristics such as curbs, need to be carefully processed, otherwise they can generate misleading results.

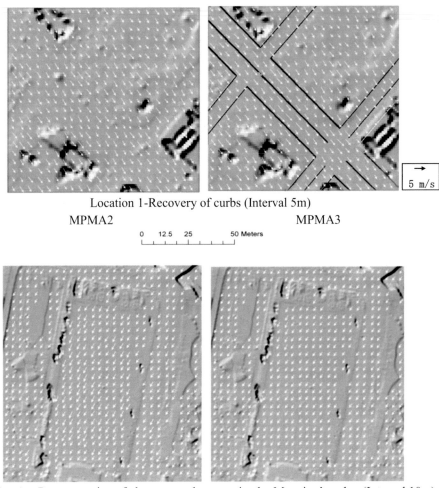

Location 1-Recovery of curbs (Interval 5m)

MPMA2   MPMA3

0   12.5   25   50 Meters

Location A – Representation of close-to-earth vegetation by Manning's value (Interval 10m)

MPMA2   MPMA3

0 12.5 25   50 Meters

Figure 8.36: Comparison of velocity vectors and DTMs generated from MPMA2 (left image) and MPMA3 (right image) at two sample location.

8.6.4 Distribution of roughness coefficient (Manning's value)

Traditionally, hydrodynamic models are provided with Manning's values manually using estimates based on a visual interpretation of available land cover information for the model domain. In this study, the input of a Manning's value is based on the extraction of points during the filtering process. The distribution of the Manning roughness number used in the flood models is dominated by three main values which are 20 for the area covered with passage building, 40 for the area covered with close-to-earth vegetation and 30 for anywhere else. In relation to this, the comparison of the Manning's number used in the models developed from the MPMA2 and MPMA3 algorithms was undertaken at two typical locations (Figure 8.37). Figure 8.23 shows that the value 30 is dominant for the study area followed by 20 and 40. It is obvious that in an urban environment there is more impervious area (represent by the values 30 and 20) than pervious area (represent by the value 40). In Location A, 57% of the area ($0.053km^2$) is represented by the value 30, 30% is represented by the value 20 and 13% is represented by the value 29. In Location B, there are only two (2) Manning's values observed, which are the values 30 (62%) and 20 (38%). Close-to-earth vegetation is not present at this location.

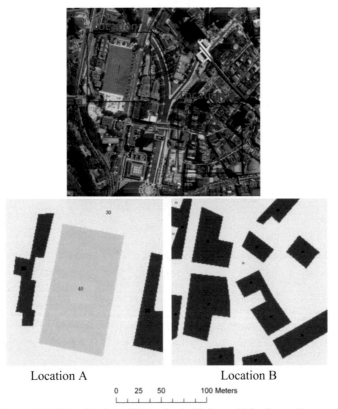

Location A   Location B

0   25   50   100 Meters

Figure 8.37: Distribution of Manning's Value within the study area

147

From all the evaluations above, it is seen that the developed algorithms show better results in term of flood depth, flood extent and flood velocity compared with the existing filtering algorithm. The combination of the filtering process, the incorporation of solid buildings and curbs in the DTM and the representation of areas with appropriate roughness coefficient (Manning's value) in the developed algorithm really contribute to better urban flood models which use DTMs that represent as close as possible the realistic urban surface.

8.7 Sensitivity Analysis

The main purpose in sensitivity analysis is to examine how sensitive the choices are to the changes in criteria weights. This is useful in situations such as where uncertainties exist in the definition of the importance of different factors. In this research, the sensitivity analysis is done to see which element in urban surface (building, elevated road, river, curbs and close-to-earth vegetation), when treated differently from the existing filtering algorithm influences the result of urban flood model in terms of flood depth and flood extent.

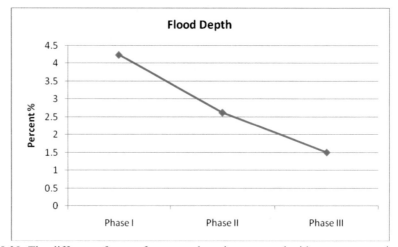

Figure 8.38: The different of errors for every phase in compared with measurement in term of flood depth

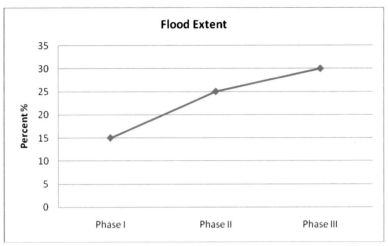

Figure 8.39: The different of errors for every phase in compared with measurement in term of flood extent

From Figure 8.38 it is seen that elements in Phase I which represent buildings have the most influence on the flood depths. This means that better handling of buildings in the case study area gives the most impact compared with the elements handled in Phase II (elevated roads, river and bridges) and Phase III (curbs and close-to-earth vegetation). This shows that for flood depth, the model is more sensitive to an appropriate representation of buildings in the urban DTM.

From Figure 8.39 the most significant differences seen in Phase II, result from elevated roads, river and bridges being handled better in the DTM using methods that combine height and slope difference with LiDAR intensity values and the interpolation of river banks. It is seen that the proposed methods incorporated in the developed algorithm contribute to a better presentation of the urban DTM. It also follows that the results for flood extent are more sensitive to the modifications done in Phase II than in Phase I and Phase III.

From the overall results, it is seen that curbs and close-to-earth (element treated in Phase III) were the least sensitive elements. Even though these elements do not seem to contribute much to flood depth and extent, the presence of curbs and the representation of close-to-earth surface roughness by 40 (Manning's value) slow down the flood flow and contribute to a more realistic diversion of the flood flow.

8.8   Discussion

Phase I: Detection and modification of buildings (MPMA1)
In Phase I, a modification was made to the existing Morphological algorithm for the buildings in

urban area. Buildings are divided into three groups which are: buildings with basements, passage buildings and solid buildings. In order to evaluate the corresponding filters, five different algorithms including MPMA were used to filter raw LiDAR data for an urban area in Kuala Lumpur, Malaysia. Such data were used to set up five different 2D floodplain models of the same area in order to investigate suitability of different algorithms for urban flood modeling applications. The model was run for two flood events (29th October 2001 and 10th June 2003). The model results were analysed and compared with recorded data in terms of flood depth, flood extent and flood velocity. From the results, it is shown that there are good correlations for all filters with observations for both flood events, but the overall analysis of results shows that MPMA has more promising capabilities then other algorithms tested in the present work. Additionally, from the sensitivity analysis, it is shown that the dataset from this phase has the most influence on the difference error in flood depth when compared with the measurements.

Phase II: Detection and modification of elevated road and bridges
This Phase describes efforts to improve one of the LiDAR filtering algorithms so that it results in safer and more accurate 1D/2D urban flood modelling work. The particular algorithm was further improved to deal with objects such as elevated road/train lines and bridges. Such geometric 'discontinuities' can play a significant role in diverting the shallow flows that are generated along roads, through fences and around buildings. The key aspects of this algorithm are: ability to deal with different kinds of buildings, ability to detect elevated road/train lines and represent them in accordance with reality, and ability to deal with bridges and riverbanks. The first aspect has been dealt with in the development of the MPMA1 algorithm, whereas the second and third aspects are original aspects of the MPMA2 algorithm.

In terms of the detection of elevated road/train lines, the data fusion method was combined with the analysis of LiDAR intensities, and those points that fall within the range of asphalt intensities were identified. This was done by using intensity, height, and slope information derived from the LiDAR data. Automatic procedures were developed within the code to remove the elevated roads/train lines and to incorporate the underneath piles. Furthermore, detection of bridges and their cross-referencing against geometric details of 1D model was also undertaken. After that, the bridges were removed from the DTM (as the river was modelled with the 1D model), and left and right banks of the river were interpolated using Kriging method.

The modelling framework applied in the present work involved building the bathymetry for 2D models using two filtering algorithms (MPMA1 and MPMA2), setting up of 2D models within MIKE21 system and coupling them with an existing 1D model (MIKE 11) which was developed in a previous study of Kuala Lumpur and carried out by DHI Water and Environment. Following this, the two sets of 1D/2D models were obtained and simulated for a rainfall event that occurred on 10th June 2003. The model results were then analysed in terms of flood depths and extent of flooded area, and compared against the observations collected at several locations.

The overall comparison of results suggests that the results of a 2D model built from MPMA2 algorithm have found to be in a closer relation to the measurements than the results of a 2D model built from MPMA1 algorithm. The difference in flood extents produced by two models was found to be in the order of 15% of the total area. In terms of the flood depth comparisons, at six out of eight locations the differences between the measurements and results from the 2D model built from MPMA2 was found to be in the range 0.17% to 19.08%. In terms of the differences between the measurements and results from the 2D model built from the MPMA1 algorithm, this was found to be in the range between 0.4% and 24.69%. However, at locations Dataran Merdeka (1) and Jalan Parlimen (5), these differences were found to be more significant for both models and they can be largely attributed to the effects of spatial variability of rainfall, lack of high resolution rainfall data and proximity of measurements taken at identified locations.

Evaluation of Phase II demonstrates that the terrain topography and the incorporation of urban features represent a factor that could make for substantial differences between the results obtained from models with different level of details and the corresponding DTMs. Furthermore, from the sensitivity analysis, it is shown that the datasets are sensitive to the difference error in flood extent when compared with the measurements. This highlights the need for a careful processing of raw LiDAR data and a cautious generation of a DTM which is aimed to be used in urban flood modelling work. Furthermore, the requirements for the description of terrain features and DTM's resolution must involve a careful consideration of the geometrical features of the area under study and the objectives of the work.

Phase III: Detection and modification of curbs and close-to-earth vegetation

From Phase III, further modification has been made to the developed filter algorithms, in order to recover the curbs in the DTM. In addition, the close-to-earth vegetation has been removed and represented by the appropriate Manning's coefficient value. By completing this phase, the development of MPMA is completed. The model simulations were performed using the data from the flood event that occurred on 10th June 2003 in order to make an evaluation of the new MPMA3 algorithm. The model results were analysed and compared with those using MPMA2 in terms of flood depths and flood extent. From the analysis of results, it can be observed that some improvement has been achieved in the context of flood depth while in terms of flood extent, the results show an increase of about 5% of the flooded area. From the overall results, it is seen that elements treated in Phase III were the least sensitive elements. Even though these elements did not seem to contribute much to flood depth and flood extent, they help to slow down the flood flow and contribute to a more realistic diversion of the flow.

The work described in Phase III demonstrates that the weakness in capturing micro features such as curbs is due to insufficient grid resolution and discontinuities present in urban environments, especially in the street network which can play a significant role in diverting and directing shallow water flows. Therefore the selection of an appropriate LiDAR filtering algorithm and the grid resolution play an important role in flood modelling work.

Chapter 9

Conclusions and Recommendations

9.1 Conclusions

While airborne laser scanning systems (LiDAR) have come a long way in recent years, the choice of appropriate data processing techniques for urban flood modelling is still being researched. Recently, several filtering algorithms have been developed and modified. Most of them run the classification process in both semi-automated and also automated ways. Nevertheless, these filtering algorithms have their own strengths and weaknesses. Sometimes the process fails to completely classify or filter out features such as vegetation or structures. When these "ruins" remain in the data set, they degrade the resulting DTM by assimilating these features into the earth's surface. The best DTM can only be obtained if the user knows the accuracy requirements for its application and how to manipulate the parameters of each filtering algorithm so that it is appropriate for the condition of the study area, whether it is an urban or forested area. Moreover, the capability of the existing filtering algorithms varies according to what type of data is being processed; whether there are high buildings, extensive forestry or other factors. Thus, the user needs to know each filter's strengths and weaknesses.

9.1.1 Development of Modified Progressive Morphological Algorithm (MPMA)

In this research, an experimental study of seven different filtering algorithms was done to assess the advantages and weaknesses of different classification algorithms in developing the most suitable DTM for Urban Flood Modelling. The major findings of the study were that:

    i.   Most of the current filtering algorithms that have been assessed face a problem in filtering urban landscape data;

    ii.   Most of the current algorithms are weak in preserving discontinuities in the bare earth (for example, road-side curbs);

    iii.   Buildings, which are the dominant objects in the urban landscape, need to be treated differently to accommodate Urban Flood models;

    iv.   The support from external data analysis (such as the data fusion method) is very useful to increase the algorithms' performance; and

153

v.   Features like curbs that are extensions of the bare earth and have a size smaller than the grid resolution, need to be detected separately and incorporated within the DTM.

In order to solve some of the above problems, a modified algorithm was developed. This algorithm begins with the existing Progressive Morphological Algorithm and modifies it to incorporate or remove particular urban features based on their needs by urban flood models. As a result, the existing PMA algorithm was modified in three phases.

Phase I incorporated *buildings with a basement, buildings with passages* and *solid buildings*. New assumptions for closed polygon buildings in vector form were introduced in order to identify different building features from the point cloud data. The novel concept of downward expansion in representing building basements was also introduced. The effect of floodwater entering and ponding in the basement of a building first before continuing to flow into the surrounding area is obviously a critical aspect, and models which do not have such a representation are unlikely to produce appropriate results. For those buildings that act as 'passage buildings', different roughness parameter values were used, where the Manning's value 25 is determined to be appropriate. By using a Manning's value less than what is applied to the overall area, the flood flow is made slower when it passes through the area. The solid buildings or objects are left in the DTM. While the concept is in contradiction with the typical existing filtering algorithm, the presence of the solid buildings or objects in the DTM contributes to a better and more realistic way of diverting the flood flow during urban flood events.

A further modification was made in Phase II to the filter algorithms. In this phase, modifications were made to remove elevated roads and bridges from the DTM. The separation process is performed by selecting those points that have intensity values in the LiDAR data between acceptable ranges for the type of road material being detected, which in this research is asphalt/bitumen. The intensity value of 'asphalt' is approximately 10~20%, and by using this intensity value, roads can be identified from the points cloud. Even though the intensity values returned by the scanning unit were noisy, sections of road material were typically uniform for road/rail lines, and as such they can be distinguished relatively easily. By searching for a particular intensity range, it is possible to extract the points that refer to road/rail lines on the elevated surface. The almost complete removal of elevated highways in the case study area, resulting from the combination of height and slope difference with LiDAR intensity value adopted in MPMA, really contributes to a more realistic flow movement during the flood event. The realistic flow movement will, of course, impact on the results of the urban flood models, especially in terms of flood depth, flood extent and flow velocity. The incorporation of piles beneath such roads and bridges was also introduced. The new algorithm allows for the incorporation of piles underneath elevated road/train lines, if the necessary information is available. Within the algorithm, this process is made optional so that if the information about the dimension of piles, the distance between them, and their location is not available, the filtering and classification process can still be performed. In this option, points that have been identified as elevated road/train lines are converted into vector form as lines before they are removed from the DTM. Using these lines as a basis, the pile shapes are incorporated in the vector form, based on locally surveyed information. The information needed for this reconstruction process includes the piles' height, width, length, and the distance between the piles. The algorithm also permits the piles to be placed in the middle or in a perpendicular position to the line. Once the vector reconstruction process is complete, the created polygons are converted into a grid and then merged back with the DTM. In addition, river surfaces, which were originally captured from the LiDAR point cloud, are removed so that the channel banks can be interpolated using the left and right bank elevations. The process of detecting and removing bridges and interpolating the river surface has been implemented within the new algorithm using the data fusion concept. The river polygon data are laid over the points cloud data using a buffer of 5m on both sides of the river, and the points within the river polygon are then extracted. From the selected points, the points with intensity values that correspond to asphalt are used to identify the location of bridges and to remove them from the DTM. Information about the location of bridges is cross-referenced with the geometry of 1D model to ensure that all bridges (and culverts) are correctly incorporated within the 1D model. The river banks are then interpolated between the left and right bank elevation values by applying the Kriging interpolation method, and the resulting DTM is used in setting up the 2D model and coupling it with the 1D model.

In Phase III, a modification was made to recover the discontinuity introduced by curbs. Vectorization and a sub-grid approach were introduced. In this research, a semi-automatic approach for the detection of a road is applied. This approach combines the vector data and raw LiDAR point cloud without converting it into any other raster format. The idea behind this approach is that raw points preserve the originality of the surface before they are converted to any other format. It works based on assumptions that:

     i.   Curbs lie much closer to the bare earth; and

    ii.   Curbs usually create continuous lines.

A high resolution grid is often needed to divide the data set into small tiles geographically. This may result in a poor performance when dealing with a large study area. To address this issue, in this research a sub-grid was adopted to create a finer set of grid tiles in certain important areas and a coarser grid was set for a quick representation of large areas. A grid-merging algorithm was proposed to produce seamless landscape scenes consisting of multiple tiles for different grid layers. In this phase, close-to-earth vegetation was also detected and represented by appropriate Manning's roughness coefficient values. A general problem in modelling surface runoff in urban areas is the lack of information about the characteristics and distribution of urban soils. Another major disadvantage is the lack of knowledge about the spatial distribution of land cover, such as grass and bush, which highly influences the amount of surface runoff. Whereas information on urban soils is very difficult to obtain and usually associated with time-consuming field work, the degree of surface sealing is estimated using the distribution of Manning's roughness coefficient.

155

9.1.2 Evaluation of MPMA using the Urban Flood Model

The evaluation of the proposed filtering algorithm MPMA is done by applying the DTM from the filtering results to flood models. The results from the flood models have been compared in terms of four difference aspects; namely flood depth, flood extent, flood velocity and distribution of surface roughness.

In order to make the evaluations of existing filters with the MPMA1, five different algorithms including MPMA1 were used to filter the raw LiDAR data for the study area located in an urban region of Kuala Lumpur, Malaysia. This area is contained within the Klang river basin. Such data was used to set up five different 2D floodplain models of the same area in order to investigate the suitability of the different algorithms for urban flood modelling application. The model was run for two flood events: 29th October 2001 and 10th June 2003. From the results, it is shown that there are good correlations of the predicted with the observed water levels for both flood events using the DTMs produced with each filter. However, the overall analysis of the results shows that MPMA1 has more promising capabilities than the other algorithms tested in the present work. Additionally, from the sensitivity analysis, it is shown that the dataset from this phase (Phase I) has the most influence on the difference error in the flood depths when compared with the measurements.

In order to evaluate the work done in MPMA2 (Phase II), six sets of 1D/2D models were obtained and simulated for a rainfall event that occurred on 10th June 2003. The overall comparison of results suggests that the results of a 2D model built with the MPMA2 algorithm have better agreement with the measurements than the results of a 2D model built from MPMA1 algorithm. The difference in flood extents produced by the two models was found to be in the order of 15%. In terms of the flood depth comparisons, at six out of eight locations, the differences between the measurements and results from the 2D model built from MPMA2 were found to be in the range 0.17% to 19.08%. In terms of the differences between the measurements and results from the 2D model built with the MPMA1 algorithm, these were found to be in the range 0.4% to 24.69%. Evaluation of the results from this phase (Phase II) demonstrates that the terrain topography and the incorporation of urban features represent a factor that could make a substantial difference between the results obtained from models with different levels of detail and the corresponding DTMs. Furthermore, from the sensitivity analysis, it is shown that the dataset is sensitive to the difference errors in flood extent when compared with the measurements. This highlights the need for a careful processing of the raw LiDAR data and a cautious generation of a DTM which is aimed to be used in urban flood modelling applications. Furthermore, the requirements for an accurate description of terrain features and the DTM's resolution must involve careful consideration of the geometrical features of the area under study and the objectives of the work.

In Phase III the model simulations were performed using data from the flood event that occurred on 10th June 2003. From the analysis of results, it can be observed that some improvement has been achieved in the prediction of flood depths while in terms of the flood extent, the results are almost the same for both models. From the overall results, it is seen that the elements treated in Phase III were the least sensitive in affecting the model results. Even though these elements did not seem to contribute much to flood depth and flood extent, if present, they did help to slow down the flood flow and contribute to a more realistic diversion of flood flows. The work described in this phase (Phase III) demonstrates that the weakness in capturing micro features such as curbs due to an insufficient grid resolution and discontinuities present in the urban environment - especially in defining the street network - can play a significant role in diverting and directing shallow water flows. Therefore, the selection of an appropriate LiDAR filtering algorithm and the grid resolution play an important role in flood modelling applications.

9.1.3 Contribution to knowledge

The success of this research is in the development of a proposed filtering algorithm MPMA that contributes to the field of Airborne Laser Scanning (LiDAR) and urban flooding in several ways. The development of an MPMA, designed specifically for urban flood modeling, can provide a high-resolution topography from aerial LiDAR data to facilitate more accurate urban flood simulations and forecasts.

The novel approach of attaching depth cavities for *buildings with basements* can fill the gap left by the existing filtering algorithm in which the standard approach for dealing with buildings is either to completely remove them from the DTM or to rebuild them as solid objects. The approach taken here is to attach the *building with basement,* with the basement properties. This is achieved by lowering the area of the building to a specific height below ground level. This condition produces the so called 'retention pond imitation', in which flooding can inundate the building area first before it floods the surrounding area. This leads to a more realistic flow movement.

Detection of elevated roads and train lines using a combination of slope and intensity in MPMA helps the filtering process in removing these elements. Most of the current filtering algorithms do not distinguish between roads and elevated roads. To produce the most accurate flood model result, it is essential to detect and remove elevated roads from the DTM. This is because their long linear form acts as a wall which diverts the flood flow, even though in real situations the water can pass through it. In this research, elevated roads are separated from other objects using the intensity value of asphalt, which is the main material used in road construction within the study area.

Detection of curbs and the recovery of a curb's discontinuity contribute to more realistic flow movements in urban flood events. This is because curbs can be found everywhere in the urban environment. In this research, a 1m x 1m grid is used as an input to represent overall urban area.

157

As the urban flood model cannot represent curbs perfectly - and making the discontinuity the main problem - the proposed filtering algorithm uses the advantage of a higher resolution grid to recover the discontinuity of the curbs by adding them to the area where discontinuities occur as a sub-grid.

MPMA assigns a roughness coefficient to the area containing close-to-earth vegetation because not all vegetation is removed from the DTM by the filtering algorithm due to limitations of the algorithm - especially for vegetation that lies close to the earth like grass and shrub. The area with this land use is detected and assigned with an appropriate roughness coefficient. The difference in roughness coefficient leads to a slower flow movement, which represents more accurately the flow as it happens in a real urban flood event.

9.2   Recommendations

High-resolution topography from aerial LiDAR data is one of the most powerful new tools for studying the earth's surface and overlying vegetation. These datasets have great utility for a wide variety of earth science and engineering applications, such as urban flood modelling. Access to raw LiDAR point cloud data is important for scientific research so that users can take full advantage of the information contained within these massive datasets. The size of these publicly available datasets presents a significant challenge to distribute them to users. Once users gain access to the data, they face significant software, hardware and expertise requirements to generate digital elevation models and perform analyses of these models. In this research, an improved filtering algorithm for urban flood modeling has been proposed. The algorithm is particularly designed to be extendible. Because of insufficient time, it was not possible to fully extend and test the algorithms, and therefore, some aspects of the research require further attention.

1.   The improved algorithm does not work as well in areas that differ from an urban landscape. Further investigations of other features may help to improve the algorithm's application.

In this research, three elements that are synonymous with an urban environment (buildings, elevated roads and curbs) have been selected, and the methods which are specifically developed to handle these elements in the resulting DTM have been adopted. In a different landscape where these elements are not essential, such as in a forest, agricultural or rural (village) landscape, the approaches adopted here in the development of MPMA will not have been utilized. Further investigation of other features such as the canopy formed by trees, scrub and bushes, and natural tracks, with their suitable treatment, would help to make the filtering algorithm more robust.

2. The classification between buildings and vegetation posed a considerable difficulty that has still not been properly resolved.

The classification of buildings and vegetation posed one of the greatest problems. These problems mainly arose from the merging of buildings and vegetation into the same categories. It was found that the combination of the capability of the predecessor PMA with the slope difference method can lead to the misclassification as it exists between vegetation and buildings. Because of this, after the building detection processing step in Phase I, it is necessary to visually verify the detections. A possible means of detecting this misclassification is a strongly-focused data fusion concept using accurate satellite images such as *Worldview* and *Quickbird*.

3. The detection of an elevated road using only intensity value works if the asphalt material is being used in its construction. The intensity values for other road construction materials need to be investigated further.

In this research, the elevated roads in the study area are found to use asphalt material in their construction. There is some previous research (see Chapter 4) that already estimated the LiDAR intensity's values for asphalt. The corresponding range of values was adopted in this research. Other construction material for elevated roads will not give the expected results for asphalt. Further investigation on the intensity values for other construction material used on elevated roads would be really useful for other urban areas.

4. The classification between curbs, pavements and close-to-earth vegetation can be misleading if only the LiDAR information is used. Support from external data (e.g. road map, topographical map) using the data fusion concept is needed and should be explored.

Curbs, pavements and close-to-earth vegetation in a typical filtering algorithm are treated as bare earth. This is because these elements are much closer to bare earth and can therefore lead to misclassification of the elements as, say, bare earth. The data fusion concept using the local plan or detailed design drawings is one of the ways that can be adopted to solve this problem.

5. Many of the parameters in the filtering algorithm depend on the information available. Their values may not necessarily have been optimal due to the lack of certain information.

In Phase I of the MPMA development where the categorizing of buildings is done, the information about the height of the building is crucial. This is because the categorization of buildings depends on their heights. Lack of this information will certainly lead to misclassification between buildings with basements, passage buildings and solid object buildings. For buildings with a basement, the average basement depth information is also important. This is to ensure the so-called flood retention pond will not be overestimated in terms of the depth available. In Phase II, the proposed filtering algorithm allows for incorporation of piles underneath the elevated road/train lines - again, if the information is available. The

159

information needed includes the dimension of piles, the distance between them, and their location. Even if this information is not available, the filtering and classification process can still be performed. However, the lack of this information will result in the capability of the proposed filtering algorithm not being fully utilized.

Therefore, whereas the algorithm has been tested against reference data, it is not guaranteed that the filter will always perform as expected. If the type of environment being filtered is untested, then unpredictable results can and should be expected.

Appendix A

Hydrological Modelling

Nedbør-Afstrømnings-Model (NAM Model)
NAM is a lumped, conceptual rainfall-runoff model simulating overland flow, interflow and base flow as a function of the water storage in each of four mutually interrelated storages representing the storage capacity of the catchment. The NAM method can include man-made interventions in the hydrological cycle such as irrigation and groundwater pumping into account. The NAM model represents the various components of the rainfall–runoff process by continuously accounting for the water content in four different and mutually interrelated storages where each storage represents a different physical element of the catchment.

These storages are:
- snow storage,
- surface storage,
- lower zone (root zone) storage,
- groundwater storage.

The model structure is shown in Figure A1 and a brief description of these parameters is given in Table A1.

Table A1: NAM parameters and their definitions

Umax	Maximum contents of surface storage
Lmax	Maximum contents of root zone storage
CQof	Time constant for interflow
CKIF	Overland flow coefficient
TOF	Root zone threshold value for overland flow
TIF	Root zone threshold value for inter flow
TG	Root zone threshold value for recharge
CKBF	Time constant for routing base flow
CK1,2	Time constant for routing overland flow

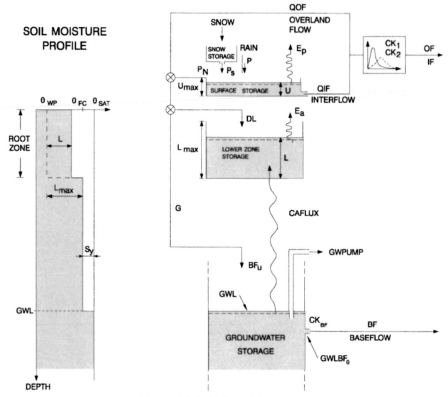

Figure A1: NAM model structure

The initial conditions required by the NAM model consist of the initial water contents in the surface and root zone storages together with the initial values of overland flow, interflow and baseflow. If the snow module is included, the initial value of the snow storage should be specified. The initial condition statement is the most important part of the model. Parameters within the model depend on the initial conditions if the simulation period is less than a year (DHI, 2004).

Time-Area Model

The time-area model of rainfall-runoff transformation is widely known as a hydrologic watershed routing technique to derive the discharge hydrograph due to a given excess rainfall hyetograph. In this technique, ignoring storage effects, the watershed is divided into a number of cells separated by isochrones. The number of cells is given by:

$$n = \frac{t_c}{\Delta t}$$
(Equation A1)

where, t_c = concentration time
        Δt = simulation time step

The histogram of consecutive contributing watershed zones from the outlet in the upstream direction is known as the time-area histogram and constitutes the basis for the excess rainfall-runoff routing. The time base of the histogram is known as the watershed time-to-equilibrium, which must be divided into a number of equal time intervals. This time interval is the travel time difference between adjacent isochrones. Using the histogram, the runoff hydrograph can be determined through convolution. The time-area technique is believed to work best for small to intermediate steep watersheds where runoff process is mainly governed by translation. In the time-area method, the influence of the shape and detailed drainage pattern of the watershed may be seen, provided that the isochrones are determined based on the watershed geomorphologic and hydraulic characteristics. Figure A2 shows the type of catchment shapes in time area-curve.

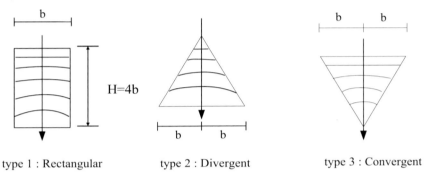

type 1 : Rectangular   type 2 : Divergent   type 3 : Convergent

Figure A2: Type of catchments shape in time area-curve (DHI, 2003)

Although many consider the TA approach a lumped model, it actually has the potential to perform as a distributed model by incorporating non-uniform excess rainfall and spatially variable watershed characteristics

Appendix B

HYDROLOGICAL MODELLING SETUP BY DHI, 2004

Modelling Approach

The hydrological modelling system applied in the flood mitigation component of this project is the rainfall runoff (RR) module of the MIKE 11 modelling package for rivers and channels. This system has been selected in order to comply with the following requirements:

i. The model shall be able to simulate to the observed peak flow characteristics of the Basin on the basis of the recorded rainfall information.

ii. Due to the rapid developments taking place in the basin the model shall have a capability to reflect the impacts from future changes in land use characteristics.

iii. The model shall be compatible with the flood forecasting system, in order to avoid doubling of calibration work and maintenance of more than one modeling system in DID.

iv. To service the environmental component of the project by supplying the hydrological inputs to water quality modeling the model shall be capable of simulating the low-flows in the main rivers over a season or longer.

v. The RR Module in the MIKE 11 Package includes a number of different hydrological models of the lumped conceptual type, and a model for modelling the rapid response from urbanised areas.

The Klang River Basin has been divided into 36 sub-catchments as illustrated in **Figure B1** and further described below. For each sub-catchment the impervious fraction was estimated based on the detailed land cover study utilising both LANDSAT images and air photographs.

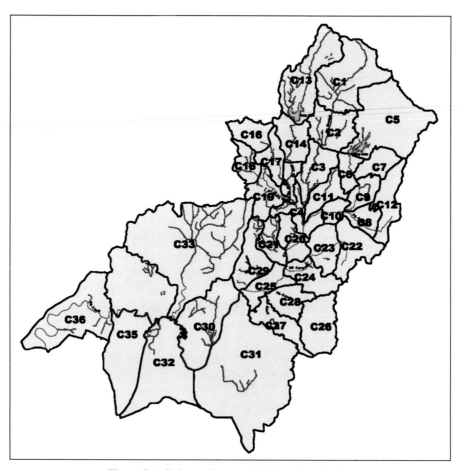

Figure B1: Sub-catchments in Klang river basin

Each individual sub-catchments have been modelled as two individual parts an impervious part, on which the urban module has been applied, and a pervious part which has been represented by the well proven NAM Rainfall-runoff model for rural areas. For each sub-catchment the total runoff from the two models are joined before entering the hydrodynamic river model.

In catchments with a pervious percentage larger than 90 the urban fraction has been neglected and the catchment has been modelled by the NAM model alone. All other catchments are divided into a pervious fraction, which is modelled by the NAM model and an impervious fraction from which the runoff is generated by the urban module.

Meteorological Input

Mean Area Rainfall

The rainfall information from the Basin originates from a network of both manual and automatic rainfall stations. All stations with records covering the last 10 years have been used for model calibration. Their locations are shown in **Figure B2**. In addition to the conventional network a smaller number of telemetric stations have been in operation in the Basin during the last one and a half years. The data from some of these stations have been used to supplement the information from the automatic stations where relevant. The data from all of the telemetric stations are being used in the flood forecasting system.

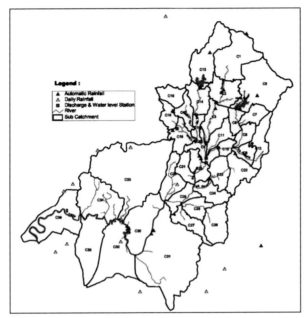

Figure B2 : Location of Manual & Automatic Rainfall Stations Used for Model Calibration

The precipitation input to the model has been calculated for each sub-catchment using MIKE11's built-in routine for mean area rainfall calculation. Because the rainfall pattern of the Basin is dominated by localised convective storms, it is important to make full use of the information from both the manual and automatic gauges when compiling the rainfall input. Large parts of the Klang Basin respond very rapidly to rainfall events and in order for the model to simulate correctly this runoff pattern, the rainfall input has to reflect the real rainfall intensities, a requirement that is fulfilled only by the automatic gauges. However, the coarser network of automatic stations does not necessarily catch the correct volume of rain, and relying on these stations alone would not necessarily lead to generation of the correct runoff volume. Therefore the rainfall volume that is being used as input to the model is the daily mean area rainfall volume for each sub-catchment calculated on the basis of the records from all manual and automatic rainfall stations using the Thiessen Method. The resulting Thiessen polygon network is shown on **Figure B3**. For each catchment the temporal distribution over the day of the calculated rainfall volume has been assumed to be equal to the observed pattern at a nearby automatic rain gauge. This method ensures the combination of all spatial information from the entire network with fine temporal resolution of the automatic stations.

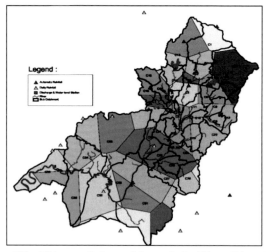

Figure B3 : Thiessen Polygon Network for Klang River Basin

Potential Evapotranspiration

In addition to the average catchment rainfall the NAM model needs monthly values of potential evapotranspiration. In this case monthly Penman estimates of average year as calculated by FAO have been applied. These values have been compared to local estimates (see **Table B1**) land found to be consistent.

Table B1 Comparison of Evaporation Estimates

Location	Method	Annual Ep (mm)
Subang Int. Airport KL	Penman	1343
Bukit Nanas	Hargreaves	1568
JPT, Ampang, Kuala Lumpur	Evaporation Pan	1469
FAO Kuala Lumpur	Penman- Monteith	1452
Klang High School	Hargreaves	1436

A water balance study of the Batu Dam catchment based on observed catchment rainfall carried out under this project indicated that the potential evapotranspiration was too large compared to the observed rainfall. However, it is not possible from the available information to pinpoint if this is due to the rainfall being too low, the potential evapotranspiration being too high or both. For practical reasons, only the potential evaporation has been corrected and a correction factor of 0.7 has been applied.

Model Calibration

Continuous moisture accounting models like the NAM model can normally be calibrated to reflect consistent water balance components such as baseflow, ground water recharge, plant available water contents, interception and superficial runoff by adjusting the parameters to flow records of around four years duration.

In the Klang basin, however, such an exercise is not straight-forward because:

The runoff series from the Gombak at the potential Damsite, which, in principle, could be used to establish model parameters for undisturbed forest areas upstream of the City, contains too many gaps and rating curve changes to allow for a reliable model calibration.

The four other gauging stations: Gombak at Tun Razak (3116433), Batu at Sentul (311434), Klang at Yap Kwan Seng (3117402) and Klang at Sulaiman bridge (3116430) are all heavily influenced by the urbanisation, that has taken place during the last decades and which has significantly altered the runoff characteristics of the basin. The aim of the modelling is to produce accurate runoff predictions for flood forecasting under real-time conditions and for evaluation of various flood mitigation scenarios under present and future conditions. Therefore only the latest years on records can be assumed to be representative for the present conditions, and form the basis for the calibration of the model.

The flow records from the recent years from several gauging stations in the area seem to be heavily influenced by the structural interventions to improve river channel conveyance that have been ongoing over the past 10 years. This work has led to unstable discharge/water level relations and many gaps in the records

Lack of firm information on reservoir extraction for water supply and return flow to the river adds further to the uncertainty of the runoff estimates.

Model Parameters for Forest Catchments

In contrast to the rest of the basin the catchments of the two dams along with the upper parts of the Gombak and Ampang Rivers are still covered with dense original forest. These catchments, which are marked C1, C2, C5, C7, C12 & C13 on **Figure B1**, also differ from the rest with respect to slope, soil types and geology. Urbanisation in these catchments is sparse and they are in the model of the present land use conditions represented solely by the NAM model.

Estimated reservoir inflows to the Batu reservoir based on reservoir water balances is regarded as the most reliable basis for a calibration of the NAM model for these catchments. The water balance has taken into account the recorded releases from the dam and the evaporation losses from the reservoir surface. The relevant authorities were contacted to obtain quantitative information on the extraction for water supply from the reservoir, however the information was not received by the project. The extraction was therefore roughly assessed as being 700 l/s or around 60,500 m³/d.

The model has been calibrated using daily time steps and the result is shown in **Figure B4**. The calibration is satisfactory both with respect to low flows and peaks. Daily observations do not suffice to calibrate the peak routing parameters in these rather small and rapidly responding catchments, and these parameters have therefore been adjusted subsequently on the basis of the records from the Batu gauging station and from Sulaiman Bridge.

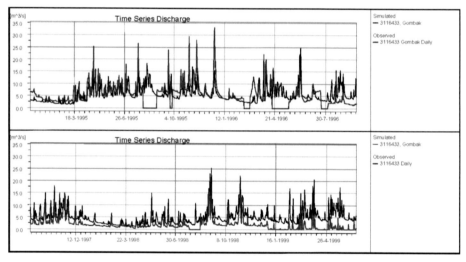

Figure 3.3.1 : Comparison of simulated inflows at the Batu Dam (black line) with inflows derived from the recorded reservoir water levels and discharge (blue line). Full series above and details of peaks below. Note : Assume extraction for water supply:700l/s.

The NAM model includes a lot of parameters to allow adjusting it to various hydrologic regimes (e.g. snow melt, ground water conditions and irrigation). However, this calibration has only been carried out on the most important parameters affecting the surface water hydrology. All other parameters (even the groundwater routing) in the Klang River model are set at the default values. The calibrated parameters are listed in **Table B2**.

The runoff coefficients for the upper catchments are very small (0.1) indicating soils with high infiltration capacity as were also found in infiltration tests. Furthermore the parameters representing the plant available water content of the soils are quite small (Umax= 30 mm) while the interception storage is big which may be explained by the rather coarse weathered granite soils and the dense vegetation cover.

The model parameters from the Batu Dam catchment have, with minor modifications, been applied to the other forested catchments.

Table B2: Calibrated Model parameters for the NAM model.

Catchm ID	Umax	Lmax	CQOF	CKIF	CK1,2	TOF	TIF
1	15	30	0.2	300	3	0.6	0
2	15	30	0.2	300	3	0.6	0
3	5	50	0.2	300	4	0.9	0
4	10	50	0.1	300	5	0	0
5	15	30	0.2	300	3	0.9	0
6	5	50	0.2	300	3	0.6	0
7	5	50	0.2	300	4	0.6	0
8	5	50	0.2	300	3	0.9	0
9	5	50	0.3	300	3	0.9	0
10	5	50	0.3	300	4	0.9	0
11	5	50	0.2	300	4	0.53	0
12	15	50	0.2	300	3	0.53	0
13	15	50	0.2	300	3	0.53	0
14	20	100	0.1	300	5	0	0
15	20	100	0.1	300	9	0.9	0
16	20	100	0.1	300	7	0.53	0
17	20	100	0.1	300	7	0.9	0
18	10	100	0.1	300	7	0.53	0
19	10	100	0.1	300	5	0.9	0
20	10	100	0.1	300	5	0.9	0
21	10	100	0.1	300	7	0.9	0

22	5	50	0.2	300	7	0.9	0
23	5	50	0.2	300	7	0.9	0
24	5	50	0.2	100	4	0.5	0
25	5	50	0.2	300	4	0.9	0
26	5	50	0.2	300	4	0.5	0
27	5	50	0.2	300	4	0.5	0
28	5	50	0.2	300	4	0.5	0
29	20	100	0.1	300	7	0.9	0
30	10	100	0.2	300	7	0.5	0
31	10	100	0.2	300	10	0.5	0
32	10	100	0.2	300	10	0.5	0
33	10	100	0.2	300	10	0.5	0
34	10	100	0.2	300	10	0.5	0
35	10	100	0.2	300	10	0.5	0
36	10	100	0.2	300	10	0.5	0

Appendix C

HYDRODYNAMIC MODEL BY DHI, 2004

The hydrodynamic model of the Klang River system, for the present condition (HD2001 Model) was successfully set-up achieving a reasonable calibration as described below.

Set-Up of the Klang River System Hydrodynamic Model

The general modelling approach and methodology for the development of the HD2001 Model (current conditions), are outlined below:

i. Cross section chainages of rivers has been measured along the centre line based on the cartographic information available and the design drawings in the river trained reaches;

ii. Boundary conditions of the model will be based on the output of the hydrological model except for the downstream boundary where tidal water level time series has to be input,

iii. Time series observation of water levels and discharges in gauging stations within the model area are a necessarily for calibration purposes. Flood level records (even spot measurements in time) are of great assistance during calibration;

iv. Hydraulic structures in the Klang river model are significant especially in terms of number of bridges in the flood prone area, and in terms of source of head losses. This variable is a key parameter during the calibration of these structures comparing to the water level difference (upstream and downstream of these bridges).

The existence of control structures in the river system with specific and defined control strategies, for instance Batu Dam and Klang gates, requires that the operation rules be collected and translated into a form that can be incorporated into the model. Operation rules are necessary for the model to be applied as a flood forecasting tool, while gate openings historical records will be used by the model during calibration and in hindcasting mode. This also applies for the Batu-Gombak diversion operation rules for all the existing / proposed control structures or for the selected flood mitigation alternatives.

River Network

Figure C1 illustrates the plan view of the Klang River hydrodynamic model. River branches are listed in Table C1 along with the upstream and downstream chainages and its confluence point to the recipient water body.

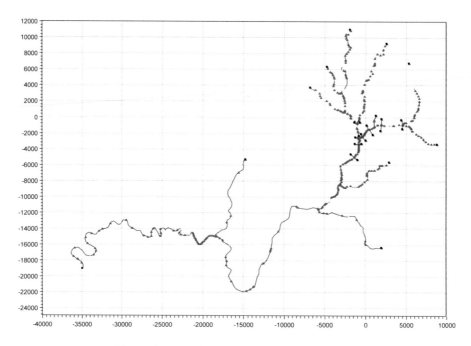

Figure C1:   Plan View of the HD2001 Model

The following rivers has been included in the current version of the Klang River system model:

Table C1   List of Rivers include in the HD Model including Chainages

River Branch Name	Upstream Chainage (m)	Downstream Chainage (m)	Confluence River Name	Chainage (m)
Klang	0	87463		
Ampang	0	5401	Klang	10100
Gombak	0	15698	Klang	16384
Batu	0	14971	Gombak	13398
Jinjang	1353	5769	Batu	10131
Kerayong	0	9394	Klang	23285
Keroh	0	5582	Batu	12527
Damansara	0	11430	Klang	53835
Kuyoh	0	11040	Klang	27716

Cross Sections

Depending on the reaches in the river system either natural river sections or design channel cross sections have been used. However in the latter case there is no as-built cross section information available. Site visits to different reaches since the start of the projects have shown that heavy siltation exists in many of these rivers. Focus was such to the flood mitigation activities specifically in the Klang, Batu and Gombak in the downtown area. Design cross sections were therefore considered no longer representative of the present condition of the rivers.

Spot cross-sections validations at different reaches e.g. the water level telemetric station (Code: 3116434) on Batu River was checked with a field cross section validation survey on 14 February, 2002. Cross sections were measured from the existing bridges from Jln Mahameru upstream to Jln Segambut, taking measurements at a few points on both sides banks, and middle of the canal. Figure C2 shows the difference between the design cross-section and the current cross-section as at 14 February 2002.

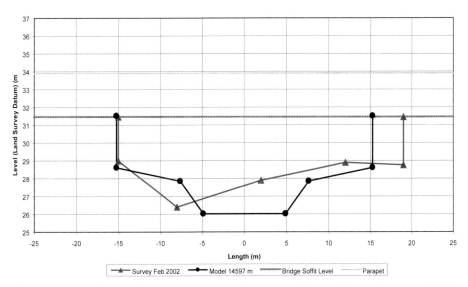

Figure C3 Batu River – Cross-section at Jln Mahameru Bridge (Model chainage 14597)

Hydraulic Structures

The HD model of the Klang River also includes several river hydraulic structures representing bridges as listed in **Error! Reference source not found.**C2. The bridges are important causes of head losses along the Klang, Batu and Gombak Rivers especially in the downtown area. For instance it is well known that the road-LRT bridge at Jln Tun Perak is an important obstacle of the flow (**Error! Reference source not found.** C4) which includes piers in the midlle of the channel as well as channel constriction created by the construction of the Putra LRT station. During medium flows it can be seen that in this constriction downstream of the bridge, formation of standing waves generates quite substantial turbulence with associated losses. This is a clear example that water levels along the rivers can be reduced in the first instance by the proper design considerations of the river flow obstructions, and by using a sound hydraulic design to minimise the head losses for different flow ranges.

Table C2 : Hydraulic structures (MIKE 11 Culvert Options) in the HD2001 Model set-up

Branch River Name	Chainage (m)	Name of the Structure	Representing in nature
SgKlang	13275	Jln Tun Razak	Bridge
SgKlang	14869	Sultan Ismail Bridge	Bridge
SgKlang	15248	Jln Dang Wangi Bridge	Bridge
SgKlang	15578	U/C	Bridge
SgKlang	15740	Medan Bunus/Jln Munshi Abdulha	Bridge
SgKlang	16160	Jln Tun Perak	Bridge
SgKlang	16442	Jln Leboh Pasar Bridge	Bridge
SgKlang	16810	Jln Tun Sambantham-Jln Cheng Lock	Bridge
SgKlang	17417	Jln Sultan Sulaiman	Bridge
SgKlang	17729	Jln Damansara/Jln Istana	Bridge
SgKlang	19760	Brickfields	Bridge
SgKlang	16175	Contraction downstream of Jln Tun Perak	
SgKlang	17185	Kinabalu Flyover	Bridge
SgKlang	19760	Brickfield	Bridge
SgKlang	21285	Federal Highway	Bridge
SgKlang	25435	Old Klang Road	Bridge
SgGombak	12998	Jalan Ipoh Bridge	Bridge
SgGombak	15419	Jln Raja / Parlimen	Bridge
SgBatu	10003	4 1/2 mile Jln Ipoh	Bridge
SgBatu	10603	4 1/4 mile Jln Ipoh	Bridge
SgBatu	10786	4th mile Jln Ipoh	Bridge
SgBatu	12951	Jambatan Jln Selvadurai	Bridge
SgBatu	13695	Jam Jln Kolam Air Empat	Bridge
SgBatu	12375	Jln Segambut	Bridge
SgBatu	13300	Jln Kasipillay	Bridge
SgBatu	13425	2 ½ Mile Jln Ipoh Railway	Bridge
SgBatu	14600	Mahameru	Bridge
CELL K06	250	Drainage K06	Drainage of associated flood area
CELL K05	250	Drainage K05	Drainage of associated flood area
CELL B01	100	Tun Razak Culvert	Existing Culvert in Bunus river at Jln Tun Razak

176

Branch River Name	Chainage (m)	Name of the Structure	Representing in nature
CELL B02	350	New Bunus	New Bunus river diversion
CELL B03	250	Old Bunus	Bunus original discharge point downtown area
CELL K01	250		Drainage of associated flood area
CELL K03	250		Drainage of associated flood area
CELL K04	250		Drainage of associated flood area
CELL K07	250		Drainage of associated flood area
CELL K08	250		Drainage of associated flood area
CELL K09	250		Drainage of associated flood area
CELL K10	250		Drainage of associated flood area
CELL G03	250		Drainage of associated flood area
CELL G04	250		Drainage of associated flood area
CELL G05	250		Drainage of associated flood area
CELL G06	250		Drainage of associated flood area

Figure C4: Klang River – Jln Tun Perak Bridge (view from left bank from upstream)

Longitudinal profiles of the different rivers included in the model are presented from **Figure C5** to **Figure C12**. The longitudinal profiles show 3 lines: a) bottom bed level, b) right margin top level and c) left margin top level. Lines b) and c) have been set to the flood level in both margins

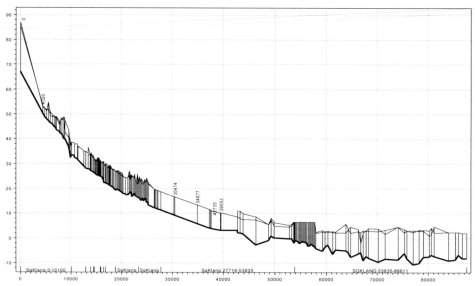

Figure C4   Klang River – Longitudinal profile HD2001 Model

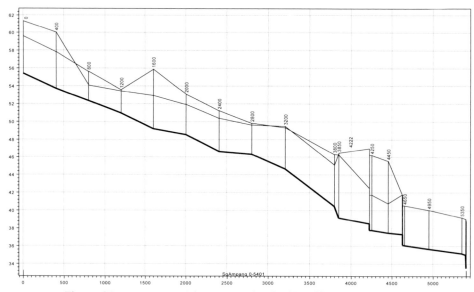

Figure C5   Ampang River – Longitudinal Profile HD2001 Model.

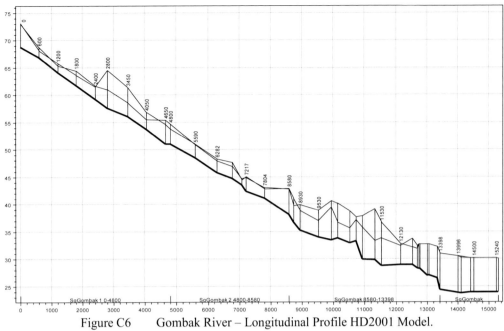

Figure C6   Gombak River – Longitudinal Profile HD2001 Model.

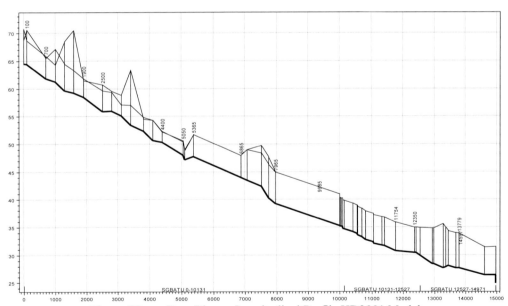

Figure C7   Batu River – Longitudinal Profile HD2001 Model.

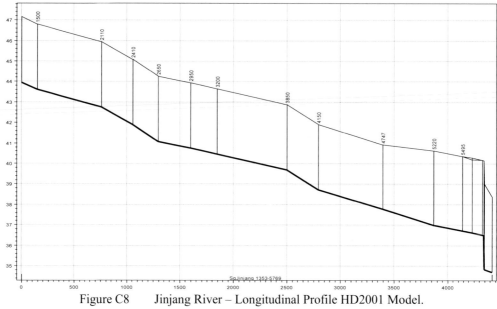

Figure C8   Jinjang River – Longitudinal Profile HD2001 Model.

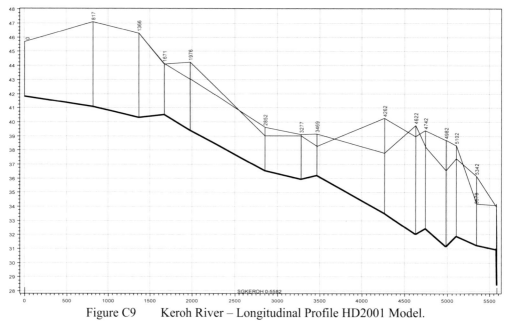

Figure C9   Keroh River – Longitudinal Profile HD2001 Model.

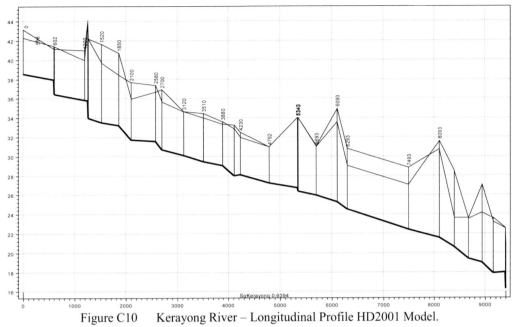

Figure C10   Kerayong River – Longitudinal Profile HD2001 Model.

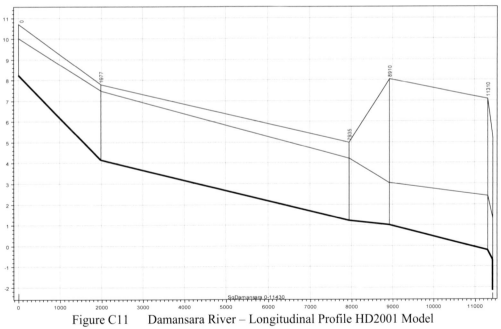

Figure C11   Damansara River – Longitudinal Profile HD2001 Model

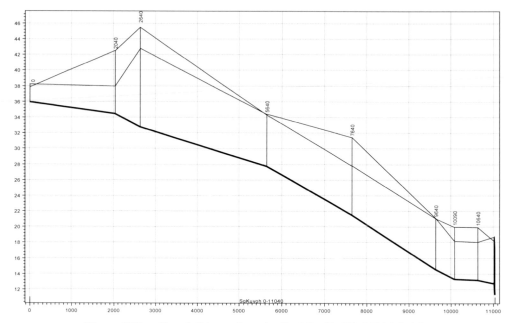

Figure C12　Kuyoh River – Longitudinal Profile HD2001 Model

Flood Plains

Within MIKE 11, this is accomplished by connecting the separated off-stream storage area, or flood cell, to the river via a long weir, represented in MIKE11 as a link channel. Flow between the river and the flood cell is then simulated in the model as flow over the weir, the crest level of which is set to the general level of the levee bank. Flood cells are storage elements only, with flow taking place to and from the river. These exchanges are dependent on water levels each side of the weir and the weir geometry only, i.e. inertial effects are neglected. In the Klang river hydrodynamic model the following flood cells has been implemented in the Klang and Gombak Rivers as listed in Table C3. In the column named "Downstream Connection" it is indicated to which water body the flood cell is connected. The area allocated to these flood cells are indicated in the Figure C12.

Table C3　Flood Plains/Flood Cell modelled in the HD2001 model

Branch Name	From (m)	To (m)	Type of Channel in MIKE 11	Upstream Connection		Downstream Connection	
Cell B01	0	200	Link Channel			CELL B01	200
Cell B02	0	700	Link Channel	CELL B01	200	SgKlang	13776
Cell B03	0	500	Link Channel			SgKlang	15733
Cell PS1	0	500	Link Channel			SgKlang	14546
Cell K01	0	500	Link Channel			SgKlang	10100

Branch Name	From (m)	To (m)	Type of Channel in MIKE 11	Upstream Connection		Downstream Connection	
Cell K03	0	500	Link Channel			SgKlang	10100
Cell K02	0	500	Link Channel			SgKlang	12800
Cell K04	0	500	Link Channel			SgKlang	12800
Cell K05	0	500	Link Channel			SgKlang	14344
Cell K06	0	500	Link Channel			SgKlang	15688
Cell K07	0	500	Link Channel			SgKlang	16695
Cell K08	0	500	Link Channel			SgKlang	16695
Cell K09	0	500	Link Channel			SgKlang	18679
Cell K10	0	500	Link Channel			SgKlang	18679
Cell G03	0	500	Link Channel			SgGombak	13398
Cell G04	0	500	Link Channel			SgGombak	13398
Cell G05	0	500	Link Channel			SgGombak	15299
Cell G06	0	500	Link Channel			SgGombak	15299

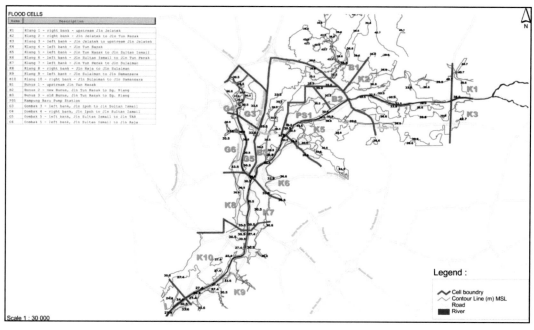

Figure C13.: MIKE 11 HD Model Flood Cells in City Area

183

Boundary conditions

The boundary conditions are the driving forces of the model. In the case of the Klang River HD model the link to the hydrological model results, as already mentioned, is the main input to the model. The external model boundaries are listed in Table C4.

Table C4 : External Boundary Conditions of the HD2001 Model

Branch River name	Chainage (m)	Boundary Conditions Type	Assign condition
SgKlang	0.00	Discharge	Klang Gates Releases
SgKlang	87463	Water Level	Tidal water levels
SgAmpang	0.00	Discharge	0 m3/s
SgGombak 1	0.00	Discharge	0 m3/s
SgBatu	0.00	Discharge	Batu Dam releases
SgJinjang	0.00	Discharge	0 m3/s
SgKerayong	0.00	Discharge	0 m3/s
SgKeroh	0.00	Discharge	0 m3/s
SgDamansara	0.00	Discharge	0 m3/s
SgKuyoh	0.00	Discharge	0 m3/s
CELL B01, B03	0.00	Discharge	0 m3/s
CELL K01 – K10	0.00	Discharge	0 m3/s
CELL PS1	0.00	Discharge	0 m3/s
CELL G03 – G06	0.00	Discharge	0 m3/s

As downstream boundary conditions, the tidal water levels generated using the 68 harmonic tidal constituents were assigned. The rest of the external boundaries are closed boundary conditions time series (Q= 0 m^3/s).

Runoff Links to the Hydrodynamic Model

The hydrological model provided the runoff to the hydrodynamic model and forms the link on how each of the catchments distribute the runoff as point values (for instance at the starting model point, chainage 0m, of Gombak, Jinjang, Ampang, Keroh and Kerayong Rivers) or distributed along each river branch as described in **Table C5**.

Table C5   Hydrological model inflow to the Hydraulic Model HD2001

Catchment Name	Catchment Area	Branch Name	Upstream Ch	Downstream Ch
C1	56.942	SgGombak 1	0	0
C2	29.27	SgGombak 1	0	2806
C3	3.769	SgGombak 1	2806	4800
C3	7.146	SgGombak 2	4800	8580
C3	13.456	SgGombak	8580	12730

184

Catchment Name	Catchment Area	Branch Name	Upstream Ch	Downstream Ch
C4	2.53	Cell K05	500	500
C4	1.392	Cell K06	500	500
C4	1.113	Cell K07	500	500
C4	0.803	Cell K08	500	500
C4	0.46	Cell PS1	0	0
C4	0.504	Cell G05	500	500
C4	2.329	Cell G06	500	500
C4	0.325	Cell G04	500	500
C4	0.514	Cell G03	500	500
C4	6.829	SgBatu	13418	14249
C4	0.119	Cell B03	0	0
C4	0.215	Cell B02	0	0
C6	2.493	SgKlang	4120	5286
C6	8.81	SgKlang	0	4120
C7	6.217	SgKlang	4120	4557
C7	9.204	SgKlang	3473	4120
C8	12.746	SgAmpang	0	5401
C9	13.493	SgKlang	5286	10100
C10	1.929	Cell K01	500	500
C10	3.116	Cell K02	500	500
C10	0.812	Cell K03	500	500
C10	2.794	Cell K04	500	500
C11	14.464	Cell B01	0	0
C11	2.437	Cell B03	0	0
C12	24.602	SgAmpang	0	0
C14	22.045	SgBatu	0	7280
C15	9.22	SgBatu	7280	13418
C16	16.953	SgJinjang	1353	1353
C17	8.685	SgJinjang	1353	1353
C17	6.97	SgJinjang	1353	5769
C18	13.517	SgKeroh	0	0
C19	21.006	SgKeroh	0	5582
C20	6.108	SgKlang	21285	22352
C20	8.177	Cell K10	500	500
C20	3.452	Cell K09	500	500
C21	11.371	SgKlang	21285	22352
C21	3.674	SgKlang	20746	21285
C22	27.871	SgKerayong	0	0
C23	21.545	SgKerayong	0	3918
C24	22.032	SgKerayong	3918	9394
C25	14.249	SgKlang	25453	30445
C25	3.619	SgKlang	24185	25435
C26	32.942	SgKuyoh	0	4990
C27	21.033	SgKuyoh	8850	8850
C28	18.501	SgKuyoh	4990	11040
C29	26.811	SgKlang	30445	31120
C30	53.87	SgKlang	31120	53835
C31	155.717	SgKlang	31120	43550

Catchment Name	Catchment Area	Branch Name	Upstream Ch	Downstream Ch
C32	68.444	SgKlang	43550	56400
C33	113	SgDamansara	0	0
C33	37.2	SgDamansara	0	11430
C34	61.903	SgKlang	53835	64700
C35	53.459	SgKlang	56400	64700
C36	65.99	SgKlang	64700	87463
C5	75.6	SgKlang	0	0

REFERENCES

Adelson, E. H. Anderson, C. H., Bergen, J. R., Burt, P. J. and Ogden, J. M., 1984, Pyramid methods in image processing. RCA Engineer, Vol. 29, No. 6. (1984), pp. 33-41

Ahmad, S., Simonovic, S.P., 2006, An Intelligent Decision Support System for Management of Floods. Water Resources Management Journal 20(3), 391–410

Abbott, M.B. & Vojinovic, Z. 2009 Applications of numerical modelling in hydroinformatics. J. Hydroinf. 11(3-4), 308–319.

Apirumanekul C. (2001). Modelling of Urban Flooding in Dhaka City. Master Thesis, No.WM 00-13, Asian Institute of Technology, Thailand

Arcement, G.J. Jr. & Schneider, V.R. 1984 Guide for Selecting Manning's Roughness Coefficients for Natural Channels and Floodplains. U.S. Department of Transportation, Federal Highway Administration, Report FHWA-TS 84, 204.

Axelsson, P. 1999 Processing of laser scanner data – algorithms and applications, ISPRS J. Photogrammetric. Remote Sens. 54(2-3). 138–147.

Baltsavias, E.P., 1999. Airborne laser scanning: Basic relations and formulas. ISPRS Journal of Photogrammetry and Remote Sensing, 54, pp.199-214.

Barkau, R.L., 1997. UNET One-Dimensional Unsteady Flow Through a Full Network of Open Channels User's Manual. US Army Corps of Engineers, Hydrologic Engineering Center, Davis.

Bartels, M. and Wei, H., Segmentation of LIDAR data using measures of distribution. International Archives of Photogrammetry Remote Sensing and Spatial Information Sciences (2006). Volume: 36, Issue: 7, Pages: 426~31

Bertrand, G., 1995. "A Parallel Thinning Algorithm for Medial Surfaces," Pattern Recognition Letters, Vol. 16, No. 9, 979-986.

Bertrand, G., 1994. "Simple Points, Topological Numbers, and Geodesic Neighborhoods in Cubic Grids," Pattern Recognition Letters, Vol. 15, No. 10, 1003-1011.

Brady, M.D. and D.R. Ford, 1990. Intelligent Data Reduction (IDARE) NAG8-642 Final Report, National Aeronautics and Space Administration George C. Marshall Space Flight Center, Huntsville, Alabama, Microfiche.

Brenner, C., Vosselman, G and Sithole, G. 2006. Presentation at the International Summer School "Digital Recording and 3D Modeling", Aghios Nikolaos, Crete, Greece, April 2006.

Brown, R.R., 2005, Impediments to Integrated Urban Stormwater Management: The Need for Institutional Reform, DOI: 10.1007/s00267-004-0217-4

Bryan, E., Christophe, V., Michael, R., Christian, P and Heiko, S., 2009, A Modelling Approach to Support the Management of Flood and Pollution Risks for Extreme Events in Urban Stormwater Drainage Systems, 4th SWITH Scientific Meeting, Delft Netherland, 4th-7th October 2009.

Boonya-aroonnet S., Maksimović Č., Prodanović D. and Djordjević S. (2007). Urban pluvial flooding: Development of GIS based pathways model for surface flooding and interface with surcharged sewer model. Proceedings of the 6th NOVATECH international conference, Lyon, France, 481-488.

Brovelli, M., M. Cannata, and U. Longoni (2004). Lidar data filtering and DTM interpolation within grass. Transactions in GIS vol. 8 (2), pp. 155 C174.

Chen, A.S., Hsu, M.H., Teng, W.H., Huang, C.J., Yeh, S.H., and Lien, W.Y., 2005, Establishing the Database of Inundation Potential in Taiwan, Natural Hazards, Springer, 2005.

Clode, S., Kootsookos, P., Rottensteiner, F., 2004. The automatic extraction of roads from Lidar data. In:ISPRS, Istanbul, Turkey, xxth congress, Commission 3, pp. 231-236.

Defina, A., 2000. Two-dimensional shallow flow equations for partially dry areas. Water Resources Research, 36(11): 3251-3264.

DHI Water & Environment (M) Sdn. Bhd., 2004, Klang River Basin Environment Improvement and Flood Mitigation Project (Stormwater Management and Road Tunnel - SMART), Final Report to Government of Malaysia.

DHI Water & Environment (M) Sdn. Bhd., 2003, MOUSE User Guide and Reference Manual. Agern Alle 11, DK-2970 Horsholm Denmark

DHI Water & Environment (M) Sdn. Bhd., 2011, Release Note 2011 . [Online] http://releasenotes.dhigroup.com/2011/MIKEFLOODrelinf.htm, accessed 28 Jan 2012

Djordjević, S., Prodanović, D., Maksimović, C., Ivetić, M. Savić, D.A. 2005 SIPSON – Simulation of Interaction between Pipe flow and Surface Overland flow in Networks, Water Science & Technology, IWA,52(5), 275-283.

Evans, B. 2008. Automated bridge detection in DEMs via LiDAR data sources for urban flood modelling. Proc 11th Int.Conf. on Urban Drainage. August 2008 Edinburgh, IWA London, UK. ISBN 978 1899796 212

Elmqvist, M. 2002 Ground surface estimation from airborne laser scanner data using active shape models. IAPRS XXXIV, 114–118.

Evans, B., 2008, Automated bridge detection in DEMs via LiDAR data sources for urban flood modeling, 11th International Conference on Urban Drainage, Edinburgh, Scotland, UK

Fairfield, J., and P. Leymarie. 1991. Drainage Networks from Grid Digital Elevation Models. Water Resources Research, 30(6):1681-1692.

FLI-MAP 400 Specifications, 2010, John Chance Land Surveys, Inc. Research Centre, [Online] http://www.flimap.com/site47.php

Garcia-Navarro P. and Brufau P., 2006, Numerical methods for the shallow water equations: 2D approach, in Donald W Knight and Asaad Y. Shamseldin (Eds.), River basin modeling for flood risk mitigation, (pp. 409-428), Taylor & Francis (2006).

Geist, T. and Stötter, J., 2002. First Results On Airborne Laser Scanning Technology as a Tool For The Quantification Of Glacier Mass Balance. Proceedings of EARSeL-LISSIG-Workshop Observing our Cryosphere from Space, Bern, March 11 – 13

Hatger, C., Brenner, C., 2003. Extraction of road geometry parameters form laser scanning and existing databases, International Archives of the Photogrammetry, Remote Sensing and Spatial Information Sciences, Vol. XXIV, Part 3/W13

Haile A.T. and Rientjes T.H.M., 2005, Effects of LiDAR DEM Resolution in Flood Modelling : A Model Sensitivity Study for the City of Tegucigalpa, Honduras, ISPRS WGIII/3, III/4, v/3 Workshop "Laser scanning 2005", Enschede, The Netherlands, September 2005

Hinz S., A. Baumgartner, 2003. Automatic extraction of urban road networks from multi-view aerial imagery, International Journal of Photogrammetry & Remote Sensing, 58: 83-98.

Horritt, M.S., 2000. Calibration and validation of a 2-dimensional finite element flood flow model using satellite radar imagery. Water Resources Research, 36(11): 3279-3291

Horritt, M.S., Bates, P.D., 2001. Effects of spatial resolution on a raster based model of flood flow. Journal of Hydrology 253 (1–4),239–249.

Houzelle, S., Giraudon, G., 1992. Automatic feature extraction and localization using data fusion of radar and infrared images. AGARD, Radio location Techniques (SEE N93-23598 08-32). 9 p. Provided by the NASA Astrophysics Data System.

Huber, R. and Lang, K., 2001, Road Extraction from High Resolution Airborne SAR using operator Fusion. In: Proceeding of the International Geoscience and Remote Sensing Symposium

Hunter, N.M., Bates, P.D., Horritt, M.S., and Wilson, M.D., 2007, Simplified spatially-distributed models for predicting flood inundation: A review, Geomorphology 90(2007) 208-225.

Hu, Y. 2003 Automated extraction of digital terrain models, roads and buildings using airborne LiDAR data. Ph.D. Thesis University of Calgary , 222 pages.

Jeong, H.S., Soo-Hee Han, Kiyun Yu and Yong-Il Kim, 2002, Assessing The Possibility of Land-Cover Classification using Lidar Intensity Data. In: IEDM Techn. Dig., p. 21.5.1-4

Jong-S. Y, Kyu-Sung Lee, Choong-Sik Woo and, Jung-Il Shin, 2006, Extraction of Bare Ground Points From Airborne Lidar Data in Mountainous Forested Area. Post-doctoral research. Department of Geoinformatic Engineering, Inha University

Konig, A., Sægrov, S., and Schilling, W. (2002). Damage assessment for urban flooding, Ninth International Conference on Urban Drainage, Portland, Oregon, USA.

Kraus, K. and N. Pfeifer (2001). Advanced DTM generation from lidar data. IAPRS vol. 34 (3W/4, WG IV/3. October 22-24, Annapolis(MD), USA), pp.23 C30.

Kubal, C., Haase, D., Meyer, V. and Scheuer, S., 2009, Integrated urban flood risk assessment – adapting a multicriteria approach to a city, Nat. Hazards Earth Syst. Sci., 9, 1881–1895

Kuiry, S.N., Sen, D. and Bates, P.D., 2010, A coupled 1D-Quasi 2D Flood Inundation Model with Unstructured Grids, J. Hydraul. Eng. 136(8), 493-506

Lamb, R., Crossley, A. & Waller, S. 2009. A fast two-dimensional floodplain inundation model. Proc. Inst. Civil Eng. – Water Manag. 162(6), 363–370. DOI: 10.1680/wama.2009.162.6.3

Leandro J., Djordjević S., Chen A.S. and Savic D. (2007). The use of multiple-linking-element for connecting surface and subsurface networks. 32 nd congress of IAHR - Harmonizing the Demands of Art and Nature in Hydraulics. Venice.

Lee, N.T., 1995. An Integrated Approache for solving urban flooding in Malaysia. Keynote address at the Seminar on Urban Drainage and Flood Mitigation organized by Institution of Engineer Malaysia, Kuala Lumpur 8-9 May 1995.

Li, S.Z., 1993. "Self-Organization of Surface Shapes," Proc. 1993 International Conference on Neural Networks, Vol. 2, October 25-29, 1993, Nagoya, Japan

Mark, B., Cheong, O., Kreveld, M. and Overmars, M., 2008, Computational Geometry: Algorithms and Applications. Springer-Verlag. ISBN 978-3-540-77973-5.

Mark, O., Weesakul, S., Apirumanekul, C., Aroonnet, S.B. & Djordjevic, S. 2004 Potential and limitations of 1D modelling of urban flooding, J. Hydrol. 299(3-4), 284–299.

Maksimovic, C., Butler, D and Graham, N., 1999, Emerging Technologies in the Water Industry. In: Water Industry Systems; Modelling and Optimization Applications, Ed. D. Savic and G. Walters, Research Studies Press, p. 39-64.

Martz, L. W., and J. Garbrecht. 1992. Numerical Definition of Drainage Network and Sub-catchment Areas from Digital Elevation Models. Computers and Geosciences, 18(6):747-761.

Mason, D.C., Cobby, D.M., Horritt, M.S. and Bates, P.D., 2003. Floodplain friction parameterization in two-dimensional river flood models using vegetation heights derived from airbourne scanning laser altimetry. Hydrological Processes, 17(9): 1711-1732.

NASA Land Processes Distributed Active Archive Center and the Joint Japan-US ASTER Science Team, 2011, ASTER Global Digital Elevation Model Version 2 – Summary of Validation Results, [Online] http://www.ersdac.or.jp/GDEM/ver2Validation/Summary_GDEM2_validation_report_final.pdf

Neal, J.C., Fewtrell, T.J., Bates, P.D. & Wright N.G. 2010 A comparison of three parallelization methods for 2D flood inundation models, Environ. Modell. Softw. 25(4), 398–411.

Neal J.C., Fewtrell T. & Trigg M. 2009 Parallelisation of storage cell flood models using OpenMP. Environ. Modell. Softw. 24(7), 872–877.

Nguyen, V. D., Bruno, M., Heiko, A., Andreas, G., Nguyen, N.H. and Delgado, J., 2008, Flood Modeling Inundated Modelling for Selected Study Area (Tam Nong, Dong Thap), The 2nd WISDOM Seminar – Can Tho, Vietnam 9-2008

Parkinson, J. and Mark, O., 2005, Urban stormwater management in developing countries, IWA Publishing.

Pfeifer, N. and P. Stadler (2001). Derivation of digital terrain models in the scop environment. Proceedings of OEEPE workshop on airborne laser scanning and interferometric SAR for detailed Detailed Digital Elevation Models. March 1-3, Stockholm, Sweden. Offcial Publication No. 40. CD-ROM., 13 pages.

Price R.K. and Vojinovic, Z., 2011, Urban Hydroinformatics: Data, Model and Decision Support for Integrated Urban Water Management, IWA Publishing

Randolph, F., 1973, Triangulated irregular network program. [Online] http://www.ecse.rpi.edu/Homepages /wrf/pmwiki/pmwiki.php/Research/TriangulatedIrregularNetwork, accessed 28 Jan 2012

Rodolfo A., 2003. Modeling highlights weaknesses in Buenos Aires drainage system. Buenos Aires, Argentina, August 12, 2003

Roggero, M., 2001. Airborne laser scanning: Clustering in raw data. IAPRS vol. 34 (3W/4, WG IV/3. October 22-24, Annapolis(MD), USA), pp. 227¨C232

Schanze, J., 2006, Flood risk management – a basic framework, in: Flood Risk Management - Hazards, Vulnerability and Mitigation Measures, edited by: Schanze, J., Zeman, E., and Marsalek, J., Springer, 149–167,

Schmitt T.G., Thomas M. and Ettrich N. (2004). Analysis and modeling of flooding in urban drainage systems. J. Hydrology, 299(3-4), 300-311

Schumann G., Matgen P., Cutler M.E.J., Black A., Hoffman L. and Pfister L., (2007) Comparison of Remotely Sensed Water Stages from LiDAR, Topographic Contours and SRTM, ISPRS Journal of Photogrammetry and Remote Sensing, doi 10.1016/j.isprsjprs.2007.09.004

Schubert, J.E., Brett F.S., Martin J. S. and Nigel G. W., 2008, Unstructured mesh generation and landcover-based resistance for hydrodynamic modeling of urban flooding. Advances in Water Resources 31 (2008) 1603–1621

Sithole, G. and G. Vosselman, 2002. Filtering strategy: working towards reliability. Proceedings of the Photogrammetric Computer Vision, ISPRS Commission III, Symposium 2002 September 9-13, 2002 vol. 34 (3A), pp. 330¨C335.

Sithole, G., 2005. Segmentation and Classification of Airborne Laser Scanner Data. Publications on Geodesy of the Netherlands Commission of Geodesy, 59, Delft, pp. 93-118

Sithole and Vosselman, 2006, Bridge detection in airborne laser scanner data, ISPRS Journal of Photogrammetry & Remote Sensing, 61, pp.33-46.

Schumann, G., Matgen, P., Cutler, M.E.J., Black, A., Hoffman, L. & Pfister, L. 2007 Comparison of remotely sensed water stages from LiDAR, topographic contours and SRTM. ISPRS J. Photogramm. Remote Sens. doi 10.1016/j.isprsjprs.2007.09.004.

Seyoum, S.D., Vojinovic, Z. & Price, R. 2010 Urban pluvial flood modeling: Development and application. 9th Int. Conf. Hydroinformatics, HIC 2010. Tianjin, China, 7–11 September.

Sithole, G. & Vosselman, G. 2004 Experimental comparison of filter algorithm for bare earth extraction from airborne laser scanning point clouds. ISPRS J. Photogramm. Remote Sens. 56, 85–101.

Sohn, G. and I. Dowman (2002). Terrain surface reconstruction by the use of tetrahedron model with the mdl criterion. Proceedings of the Photogrammetric Computer Vision, ISPRS Commission III, Symposium 2002 September 9-13, 2002 vol. 34 (3A), pp. 336 C344

Smith M (2009). Lessons learned in WASH Response during Urban Flood Emergencies. A Global WASH Cluster Lessons Learned Paper.

Song, C., 2005. Spectral mixture analysis for subpixel vegetation fractions in the urban environment: How to incorporate endmember variability?, Remote Sensing of Environment, 95, 248-263.

Trias-Sanz, R., Lomenie, N., 2003. Automatic bridge detection in high-resolution satellite images. Proc. 3rd International Conference on Computer Vision Systems (ICVS 03), Graz, Austria, April 1–3, pp. 172–181.

Tyrna, B. G. and Hochschild, V., 2010, Urban flash flood modelling based on soil sealing information derived from high resolution satellite data. International Interdisciplinary Conference on Predictions for Hydrology, Ecology and Water Resources Management 20-23 September 2010

USDA (1986) Urban hydrology for small watersheds. Technical Release 55, United States Department of Agriculture, Washington.

Vojinovic, Z., Salum, M.H., Seyoum, S.D., Mwalwaka, J.M., Price, R.K. & Abdullah A.F. 2010(a) Modelling urban floodplain inundation with different spatial resolution and model parameterisation. 9th Int. Conf. on Hydroinformatics, HIC 2010. Tianjin, China, 7–11 September.

Vojinovic, Z., Seyoum, S.D., Mwalwaka, J.M. & Price, R.K. 2010(b) Effects of model schematization, geometry and parameter values on urban flood modelling. Water Sci. Technol. (In Press)

Vojinovic, Z. & Tutulic, D. 2009 On the use of 1D and coupled 1D–2D modelling approaches for assessment of flood damage in urban areas. Urban Water J. 6, 183–199.

Vojinovic, Z and van Teffeelen, J., 2007, An Integrated Stormwater Management Approach for Small Islands In Tropical Climates, Urban Water Journal.

Vojinovic Z., Bonillo J., Kaushik C. and Price R. (2006). Modelling flow transitions at street junctions with 1D and 2D models. 7 th Int. Conf. on Hydroinformatics, Nice, France, Vol. I, 377-384.

Voogt, J.A. and T.R. Oke, 2003. Thermal remote sensing of urban climates. Remote Sensing of Environment 86, pp. 370-384.

Fortune,S., Voronoi diagrams and Delaunay triangulations , Handbook of Discrete and Computational Geometry, J.E. Goodman, J. O'Rourke, eds., pp. 377-388, CRC Press, New York.

Vosselman, G. 2000 Slope based filtering of laser altimetry data. Int. Arch. Photogramm. Remote Sens. Spatial Inf. Sci. 33(B3/2), 935–942.

Wang, Y., Zheng, Q., 1998. Recognition of roads and bridges in SAR images. Pattern Recognition 31 (7), 953–962.

Wang, C.K. and Tseng Y.H., 2010. DEM generation from airborne LiDAR data by an adaptive dual-directional slope filter. ISPRS Commission VII Symposium Thematic Processing, Modeling and Analysis of Remotely Sensed Data. Vienna, Austria

Wack, R. & Wimmer, A. 2002 Digital terrain models from airborne laser scanner data – a grid based approach. IAPRS XXXIV pp. 293–296. ISPRS Commission III, Symposium. 9–13 September 2002, Graz, Austria.

Wechsler.S. 2006. Uncertainties associated with digital elevation models for hydrologic applications: a review, Hydrol. Earth Syst. Sci. Discuss., 3, 2343-2384.

Wehr, A. & Lohr, U. 1999. Airborne laser scanning – an introduction and overview. ISPRS Journal of Photogrammetry and Remote Sensing 54: 68–82

Woolhouse, G., 2008, Infoworks 2D – a new dimension to flood modelling. In: WaPUG Spring Meeting, 3 June 2008, Coventry. (2008)

Xuelian, M., A slope- and elevation-based filter to remove non-ground measurements from airborne LIDAR data. Technical paper. Geography department of Texas State University, San Marcos

Xuelian M, Le Wang, and Nate Currit, 2008. Morphology-based Building Detection from Airborne Lidar Data. Photogrammetric Engineering & Remote Sensing Vol. 75, No. 4, April 2009, pp. 437–442

Yang, L., G. Xian, J.M. Klaver, B. Deal, 2003. Urban lad-cover change detection through sub-pixel imperviousness mapping using remotely sensed data, Photogrammetric Engineering & Remote Sensing, 69, 1003-1010

Yu D, Lane SN. 2006. Urban fluvial flood modelling using a two-dimensional diffusion wave treatment: 2. Development of a sub grid-scale treatment. Hydrological Processes 20: 1567–1583.

Yu D, Lane SN. 2011, Interactions between subgrid-scale resolution, feature representation and grid-scale resolution in flood inundation modeling. Hydrological Process. 25, 36–53.

Zakšek, K. and Pfeifer, N., 2004, An improved morphological filter for selecting relief points from a LIDAR point cloud in steep areas with dense vegetation. Technical Report. Delft Institute of Earth Observation and Space systems, TU Delft, The Netherlands

Zhang, K.Q., Chen, S.C., Whitman, D., Shyu, M.L., Yan, J.H. & Zhang, C.C. 2003 A progressive morphological filter for removing nonground measurements from airborne LiDAR data. IEEE Trans. Geosci. Remote Sens. 41(4), 872–882.

Zhang, K. & Cui, Z. 2007 Airborne LiDAR Data Processing and Analysis Tools ALDPAT 1.0 Software Manual. National Center for Airborne Laser Mapping, Florida International University, Miami.

Zhang, K. & Whitman, D. 2005 Comparison of three algorithms for filtering airborne LiDAR data. Photogramm. Engng Remote Sens. 71, 313–324.